Recommendations for
Reducing Emissions from the Legacy Diesel Fleet

Report from the Clean Air Act Advisory Committee

April 10, 2006

This page intentionally left blank.

Table of Contents

EXECUTIVE SUMMARY

A. Overview

Diesel engines play a vital role in key industry sectors such as goods movement, public transportation, construction, and agriculture. A unique combination of efficiency, power, reliability, and durability make diesel the technology of choice for these sectors. However, the durability of the technology does not lend itself to rapid fleet turnover and investment in new equipment that meets more stringent environmental standards.

Because of this, the full air quality benefits of the very stringent new engine emission standards in the US2007 Diesel Rule ("Heavy-Duty Engine and Vehicle Standards and Highway Diesel Fuel Sulfur Control Requirements.")[1] and the Nonroad Diesel Rule ("Clean Air Nonroad Diesel Rule.")[2] will likely take decades to achieve. Further, the regulatory authority of EPA and states to address the existing fleet of over 11 million diesel engines is rather limited. In response, EPA began the Voluntary Diesel Retrofit Program in 2000 to discuss broad initiatives to modernize and upgrade (i.e., retrofit) current engines with modern emission control equipment or to accelerate the replacement of these engines with newer ones.

Given the diversity of applications and engines, as well as significant technical and funding issues, the Clean Diesel Retrofit Work Group was formed in 2004 under the auspices of the EPA Clean Air Act Advisory Committee (CAAAC) to advise EPA on how best to expand the initiative. The Work Group consists of over forty members representing diverse stakeholders. It is organized under four main sectors by application: school buses, ports, freight, and construction. EPA determined these sectors to have the greatest need and potential for achieving emission reductions, based on the number and types of engines as well as exposed populations and predicted sector growth.

Although no complete analysis is available quantifying the benefits and costs, the positive return on retrofitting the current diesel fleet with the best available technology is likely significant. For example, when fully implemented, EPA estimates the EPA 2007 Diesel Rule impacting new engines and requiring cleaner diesel fuel will have returned $17 to society in health benefits for every dollar spent. The Nonroad Diesel Rule that was finalized in 2004 will deliver $40.

Although the overall benefit of reducing diesel emissions is significant, the investment needed to clean up the existing fleet is also quite large, perhaps in the range of $50 to $100 billion. The Work Group believes that this is not an

---

[1] Regulatory Impact Analysis: Control of Emissions of Air Pollution from Highway Heavy-Duty Engines. EPA420-R-00-010. July 2000. Available online at: http://www.epa.gov/otaq/hd-hwy.htm#regs.
[2] Final Regulatory Analysis: Control of Emissions from Nonroad Diesel Engines. EPA420-R-04-007, May 2004

insurmountable barrier and represents a small fraction, possibly as little as 5%, of the total cost of operating and maintaining the legacy fleet over a 10 year period.[3]

1. Incentives

A variety of incentives are available for reducing diesel emissions. In some cases, these can be combined and tailored for specific sectors. Income tax incentives can take the form of exemptions, deductions, and credits. Tax incentives are easy to use, but have the challenge of targeting cost-effective reductions. They are also not applicable to publicly-owned fleets and might not address users in low tax brackets. Reducing excise taxes, such as has been done with alternative fuels, might also be effective.

Grant programs are the most popular current funding program for retrofits. Grants provide funds directly to owners and operators to pay for new engines or vehicles, or equip existing fleets with retrofit equipment. They allow direct funding of equipment to the fleet owners. Examples are the Carl Moyer Program in California, the Texas Emissions Reduction Plan (TERP), and the EPA Clean School Bus USA grant program. Funding for such programs can also come from Supplemental Environmental Projects, wherein funding negotiated as part of a legal settlement might be targeted to retrofits. Grant money can also be used to set up low-interest loan programs. Grant programs can be very effective, but require more effort to implement on both the government and private sides than tax incentives.

Contract terms on public projects can also be used to provide incentives for retrofits. Contract terms can be used in multiple sectors and by any entity that pays for a service that is provided in part by a piece of diesel equipment.

2. Mandates

Government mandates are another and, potentially, very effective tool for forcing fleets to upgrade their equipment. California's Diesel Risk Reduction Plan is an example of such a state law. Although EPA has very limited authority to mandate retrofits, states can adopt provisions related to retrofits for on-road vehicles used within its borders, but will have to follow California's lead and obtain EPA waivers for nonroad engines. The Work Group cannot reach consensus on who pays for retrofits in mandatory programs (e.g., the end user or society) and decided to leave this discussion out of this report and these recommendations.

3. Technologies

Many established and emerging technologies are available to help modernize and upgrade the existing fleet and reduce emissions. In this report, "retrofit" is intended to broadly refer to a variety of approaches including engine replacement and

---

[3] Calculation based on: 45 billion gallons of diesel fuel consumed per year for highway and non-road (Transportation and Energy Databook, cta.ornl.gov/data) at a price of $2.50 per gallon. Five percent of this gross fuel consumption over 10 years is $56B.

recalibrations, the use of clean fuels, installation of exhaust aftertreatment devices, and anti-idling and other changes in operating strategies that reduce emissions. Each strategy has its own strengths and weaknesses. Engine replacements and recalibration can be effective and may result in enhanced fuel economy and lower maintenance costs, but can be expensive. Switching to cleaner fuels might be easy, as in the case of switching to lower sulfur fuel or biodiesel blends, but has air quality benefit on its own. Switching to alternative fuels, like natural gas, can be very effective, but one has to establish a new fueling and maintenance infrastructure in addition to engine and fuel system modifications. Retrofit technologies are effective in reducing PM, HC, CO and sometimes NOx, but care must be taken to appropriately match a specific retrofit technology with an in-use application. Anti-idling strategies are a winner across almost all applications and save fuel and money. Some idle-reduction strategies require infrastructure investments at truck stops, and the air quality benefits are lower compared to other strategies. For the purposes of establishing retrofit technology credibility and state air pollution credits, the California Air Resources Board and the EPA have developed retrofit technology verification procedures with reciprocity.

B. General Cross-Sector Recommendations

- The potential for cleaning up the existing fleet is significant and worth the investment.

- The goal of these programs is the deployment of the most feasible technology for a specific application and positive recognition.

- Given the diversity of applications, it is important to offer a range of funding options and incentives for maximum impact. Grants, loans, rebates, and tax incentives are common funding mechanisms across all sectors.

- Education and outreach is essential to spread the word and maximize impact.

- The EPA technology verification process should be streamlined to get more technology options into the market and increase competitiveness.

- The 2005 Energy Bill (Pub L 109-49)[4] and 2005 SAFETEA-LU (Pub L 109-59[5] included significant provisions pertaining to retrofits and funding. Full funding and implementation of these measures will result in the greatest emissions reductions from the legacy fleets.

---

[4] Official Title: *Domenici-Barton Energy Policy Act of 2005.*
[5] Official Title: *To authorize funds for Federal-aid highways, highway safety programs, and transit programs, and for other purposes.*

C. Sector-Specific Descriptions and Recommendations

1. School Buses

About 70% of the 480,000 school buses are owned and operated by the school districts.[6] Since the majority of school districts have very limited funding, and private contracts are tight for the remaining 30% of school bus transportation, all the funding of retrofits in this sector will likely have to come from grants. Because of the importance placed on children, and given that 30% of school buses are pre-1991 model year, reducing emissions from the nation's school bus fleet should be a first priority.

The school bus sector is leading the others on Federal funding due to an early start from the EPA Clean School Bus USA Program, which has awarded about $17.5 million in FY2003 and 2004 combined. These efforts have reduced emissions from about 7,400 buses through replacements, refueling, and retrofits. The use of clean diesel fuel accounts for the majority of the buses (3,969), followed by diesel oxidation catalysts (2,169). Further, an additional 23,000 buses have been cleaned using state programs and Supplemental Environmental Project (SEP) funds.

The recommendations to advance retrofits in this sector are:

- The amount of funds available for reducing emissions through government grants should be increased and its disbursement rate accelerated. Tax incentives, such as tax credits, sales and property tax exemptions, and waivers of registration fees will also have value for the privately-operated fleets.

- Incentive grants should be geographically diverse and focus on producing the greatest emission reductions for the least cost by concentrating on replacing the 2,000 pre-1977 buses first.

- Education and outreach is important, as many school districts are not aware of the grant programs.

2. Freight

Moving freight consumes about 20% of all energy in the U.S. Trucks move about 66% of freight, while rail moves 16%. The remaining 18% is moved by water, pipeline, and air. Ground freight accounts for 40% of transportation-related NO_X and 30% of PM emissions. Since two-thirds of these emissions come from trucks, the Work Group decided to focus on the truck sector. The rail and airport sectors should be examined later, but appear to be moving forward with their own unique efforts such as idle reduction, use of green switcher locomotives, and improvements to ground service equipment in the airport sector.

EPA's SmartWay Program, a market-based incentive that combines energy efficiency and environmental performance, is the focal point of the freight sector's

---

[6] This and other statistics taken from data published in the 2005 issue of *School Bus Fleet Magazine's Fact Book.*

incentives. SmartWay recognizes the unique relationship between shippers and freight carriers, and seeks to take advantage of that relationship through preferential contracting for trucking companies that are SmartWay partners. Carriers can get certified as such by employing state-of-the-art fuel efficiency technologies bundled with emission reduction technologies. In addition to the resulting economic benefits, wherein fuel savings provide a good return, shippers and carriers are also recognized through publicity, which has been shown to be of value. Examples of fuel saving technologies are idling controls, improved tires and wheels, and improved vehicle aerodynamics to reduce rolling resistance and drag, and hence energy consumption and emissions. Financial incentives to make these investments come from grant and loan programs.

Other incentives that should be considered include those that offer operational benefits such as privileged parking, lane access for loading and unloading, reduced or eliminated tolls, and expedited access to points of entry. Some technologies (e.g., auxiliary power units for idling reduction and urea storage tanks for SCR systems) are subject to the excise tax and add weight to the vehicle. Efforts to provide waivers for the excess weight of these technologies should be pursued, as these are barriers to innovation that should be removed.

A key factor in the freight sector is determining how to apportion SIP credits and/or air quality benefits derived from trucks traveling across multiple states and regions. States and localities should be able to claim emission reductions from implementing the above incentives in their SIPs. This involves developing formal air quality guidelines and procedures to claim credits. Also, clarity is needed in the tax treatment of grant funding such that companies are not discouraged from seeking grants for fear of tax implications.

In addition to the above, EPA should:

- Adopt programs to encourage the replacement of older, higher emitting vehicles with new, cleaner 2007 vehicles.

- Explore implementing loan programs, tax incentives, and labeling programs for hybrids.

- Further evaluate the benefits from the Smartway technology bundles and publicize the results. Create fuel efficiency and emission reduction thresholds for program participation.

- Work to expand the Smartway loan programs beyond Arkansas and Minnesota, and create a national capitalization/loan program.

- Encourage aggressive coordinated leadership between states, NGOs, and trade associations in order to implement programs that will require high levels of funding.

- Further work with the Department of Energy (DOE) on exploring new fuel economy and emissions reduction technologies for the legacy fleet, and with the Department of Transportation (DOT) to educate localities on how to better use

Congestion Mitigation and Air Quality (CMAQ) funding to help reduce emissions from existing vehicles.

One issue on which consensus was not reached was whether EPA should consider mobile-to-stationary source emissions trading for shippers.

3. Marine Ports

The United States is served by 185 deep-draft ports, and 30 of the largest ports are in ozone and/or PM nonattainment areas. Many other ports are located in maintenance areas that are former nonattainment areas. Some areas have air quality values that approach the standards on a regular basis. Many of these 30 ports are close to commercial and residential districts, which have become increasingly concerned about port-related emissions. This concern will grow as ports become more active to handle significant projected increases in freight and cruise ship volumes.

Sources of diesel emissions at ports include ocean-going vessels, harbor craft, local cargo-handling equipment (cranes, yard hostlers, etc.), and trucks and rail that carry goods in and out of them. Ports either operate their own equipment or lease their land on a long-term basis to private marine terminal operators, who own and operate their own cargo-handling equipment.

Based on the Port of Los Angeles emissions inventory,[7] port activities contribute the following NO_X and PM emissions:

- Cargo handling equipment constitutes approximately 10% of the regional NO_X and 12% of the regional direct $PM_{2.5}$ emissions.

- Heavy-duty trucks currently calling on major container ports emit about 23% of the regional NO_X and about 9% of the regional directly emitted $PM_{2.5}$.

- Rail contributes approximately 13% of the regional NO_X emissions and 6% of the regional directly emitted $PM_{2.5}$

- Marine vessels, including harbor craft (e.g., tugboats, towboats, and ferries) and large ocean-going vessels (e.g., container ships, tankers, and cruise ships), emit about 54% of the regional NO_X and 72% of the regional directly emitted $PM_{2.5}$.

While many port authorities and terminal operators have been pro-active in reducing emissions, many opportunities remain. At the same time, unique barriers also need to be considered. Grant funding is limited, and application deadlines might be out of sync with business cycles (i.e., port enhancement projects).

---

[7] *Port of Los Angeles Baseline Air Emissions Inventory* – 2001. July 2005. Available online: http://www.portoflosangeles.org/DOC/POLA_Final_BAEI_ExecSum.pdf

Administrative burdens of grants may also be too high to make them worth the effort. Ports are very competitive, so additional financial burdens will result in lost business if they are not universal. Homeland Security initiatives cause competition for funds, but perhaps also coordination opportunities. Currently, demonstrated emission reduction technology available for ports is limited, but expanding. Finally, each port is unique, and many experience local, state, and federal jurisdictional issues.

Flexible program design and education and outreach opportunities exist. Recommended incentives include grants, tax incentives, loan programs and rebates, contract or lease requirements, recognition programs, and regulatory credits.

Additionally, EPA should:

- Include grant programs for ports in its budgetary process, and create a model program to demonstrate to states how they can use their fee authority, similar to California's Carl Moyer Program or TERP.
- Coordinate with the DOT and Homeland Security to begin addressing air quality impacts of major infrastructural programs.
- Explore SIP credit structure related to ports.
- Adopt recognition, education, and outreach programs that specifically target ports.

4. Construction

The construction industry uses more diesel engines than any other sector—more than 2 million, almost 20% of the total—that vary in all important aspects (e.g., size, configuration, cycle, age). Thirty-one percent of these engines were manufactured before emission regulations were implemented, so the sector has a disproportionate amount of emissions—32% of all mobile NO_X and 37% of PM. Most of the equipment is used in public works and commercial projects. Many construction companies are small businesses: 92% have fewer than 20 employees,

This sector has some unique characteristics that provide challenges in designing programs to promote retrofit. Emissions and economic impacts depend on frequency and time of use, location, and application. Thus, priorities and programs within the sector need to be carefully considered, while specificity of incentives might be difficult. Emission control retrofits may represent risks, so full grants might not be as attractive in many applications. Also, income tax incentives might have limited value due to low profitability frequencies in the industry. Finally, as emission control retrofit technology is slowly coming into the sector, grant periods need to be long enough to allow the market to take advantage of the opportunities.

Contractual incentives are a powerful tool for the industry, but need to be carefully designed. Contract modifications, which will reward clean practices, can level the playing field for small companies when combined with attractive loan programs. However, participation might be limited unless they are attractive. Among construction companies is a concern that contract requirements and regulations could provide a competitive advantage to large, private sector equipment owners with sufficient capital to meet cleaner requirements, while discriminating against smaller businesses that could not afford to retrofit equipment.

EPA should:

- Identify ways to use Federal grants to leverage private funding, and state/local grants to maximize the pool of funds and benefits to equipment owners.

- Ensure adequate resources are available to administer programs or contract program administration out to other organizations.

- Develop model language for contract-based incentives by working with the industry and procurement officials.

- Model SIP credits for voluntary retrofits after the TERP program, where they are not part of the 3% SIP maximum, and are enforceable credits.

- Investigate emissions benefits from changing operational behavior, such as reducing idling and enhancing maintenance, and then establish appropriate guidelines.

- Develop tools to improve understanding of emissions inventories, retrofit costs, and commensurate benefits to improve policy decisions. Communicate these to state and local officials and use them to enhance construction project credits.

- Improve the technology verification process by: allowing reasonable extensions of technology from the highway sector to the nonroad sector; allowing conditional technology verification (with finite duration); considering reciprocity with the Swiss VERT process, perhaps on an interim basis.

D. Concluding Remarks

The Work Group spent much collective time and effort on assessing the various options for advancing emissions reductions via retrofit programs. The costs of emission reductions are significant, but the societal benefits are much larger. Few public investments show as much promise in providing these returns.

Along the way, awareness of the options increased in each sector, and sector champions were developed. So, in a way, some of the recommendations are well on their way to being implemented (e.g., education and outreach). Further, the Work Group is largely committed to taking the next steps to help EPA implement the recommendations. As a final recommendation, we are recommending that EPA keep the Retrofit Working Group active and alive, perhaps not in exactly the same form as it currently exists under the Mobile Source Technical Review Subcommittee (MSTRS), but in some form, as a mechanism to continue to promote improvements in diesel emissions reduction programs, nationwide.

I. Introduction

In 2000, EPA began the Voluntary Diesel Retrofit Program in response to the widely accepted need to reduce diesel emissions from the existing fleet of nearly 11 million diesel engines. Components of diesel exhaust can cause a multitude of health problems and negative economic impacts. EPA has designated diesel exhaust as a likely human carcinogen, causing many other health-related problems as well as environmental and economic impacts.

New diesel engines and vehicles have been subject to EPA's regulatory program of progressively more stringent emissions standards since the late 1980's for highway engines and since the early 1990's for engines used in nonroad applications. EPA's newest and most aggressive sets of standards for diesel engines and fuels will be phased in between 2007 and 2014. These standards will achieve up to a 95% reduction in pollution from new highway and nonroad diesel engines and vehicles. However, the newest standards do not apply to the 11 million engines in the "legacy fleet" that were manufactured to meet previously applicable but less stringent standards. Since these engines will remain in use for up to 30 more years, reducing pollution from these existing engines and vehicles would significantly reduce exposure to harmful diesel exhaust and help the Nation improve its air quality.

EPA has little authority to regulate existing engines. Also, these 11 million existing engines are operated in a wide variety of applications and owned by a complicated web of industries and businesses. These factors pose challenges for designing a program that will achieve the desired emissions reductions needed to protect public health.

The magnitude of the effort needed to create such a program led to the convening of the Clean Diesel and Retrofit Work Group, as part of the Federal Clean Air Act Advisory Council (CAAAC), under the auspices of its Mobile Sources Technical Review Subcommittee (MSTRS). The charge to this work group was to make recommendations to the Agency through the CAAAC process on how to best address the emissions from the legacy diesel fleet with a focus on creating voluntary incentive-based approaches.

The Work Group has defined the term "retrofit" to mean any diesel emissions reduction strategy that can be used to reduce emissions from the legacy fleet including, but not limited to, the use of after-treatment devices, engine replacement, recalibrations, cleaner diesel and alternative fuels, and anti-idling devices and operating strategies.

The forty-two members (see Appendix A) that officially make up the Clean Diesel and Retrofit Work Group represent the full range of groups with a vested interest in reducing pollution from the legacy fleet. The Work Group is co-chaired by representatives from EPA and Corning, Inc., and is further divided into four "Sector Groups": School Buses, Ports, Freight, and Construction. EPA determined that

1

these sectors have the greatest need and potential on a national basis for achieving emission reductions, based on the number and types of engines as well as exposed populations and predicted sector growth. On a more local or regional scale, other sectors, such as the agricultural sector maybe very important. The fact that this report does not deal with all sectors does not diminish the importance of controlling emissions from these other sectors and it is hoped that strategies and incentives outlined here will further than end.

Each of the selected sectors differs in terms of economic and business practices, which are keys to understanding how to motivate retrofit and other clean diesel strategies within each. The ports and construction sectors, in particular, will experience unprecedented growth over the next decade, and it is especially important to manage the emissions from these sectors to protect public health in adjacent communities.

The four sector groups were co-lead by an EPA staff member and an external party. These Sector Groups engaged additional experts in the process, widening participation in these discussions to well over 100 individuals.

This report is the culmination of the work of the Clean Diesel and Retrofit Work Group since April 2004. It provides consensus-based recommendations as well as other recommendations. Some recommendations are sector-specific; others apply more broadly. It is our hope that this report will substantially further our Nation's efforts to achieve healthy air for its citizens.

II. Background

Diesel exhaust plays a key role in the health impacts of air pollution,[8] and analyses have indicated that cleaning up diesel emissions has a significant benefit to society. For example, analysis of EPA's 2007 Heavy-Duty Highway Final Rule has determined that full implementation of the rule will return to society net benefits of $70 billion annually.[9] Similarly, the 2004 nonroad regulations will result in a net benefit of $80 billion annually.[10] EPA is in the process of fully analyzing the return to be realized through reducing emissions from the legacy fleet. The Union of Concerned Scientists (UCS) has estimated that for every dollar invested in retrofits, $9-16 dollars are returned to society.[11] The following discussion elaborates on the health and environmental considerations.

A. The Case for Reducing Diesel Emissions

Diesel engines emit small particles ($PM_{2.5}$) and gases, including air pollutants such as benzene and polycyclic organic matter (POM), which are known to be toxic above certain levels. Diesel engines also emit ozone-forming nitrogen oxides (NO_X) and hydrocarbons (HC). Therefore, reducing diesel emissions is an important public health issue and air quality concern. Some examples of vehicles and equipment operating diesel engines include trucks, school buses, transit buses, construction equipment, cargo-handling equipment, locomotives, ferries, and ships. Figure II.1 presents the contribution of NO_X, PM_{10}, and volatile organic carbon (VOC) emissions from mobile sources as compared to stationary sources. Figure II.2 presents each sector's contribution to the mobile source population. Figure II.3 presents each sector's contribution to $PM_{2.5}$ emissions, and Figure II.4 presents contributions to NO_X emissions.

---

[8] The impacts of air pollution are measured by indicators such as number of lost days of work, incidence of hospitalization and emergency room visits. Analysis is based on peer reviewed studies as described in Regulatory Impact Analyses. For a fuller discussion of health effects, see US EPA

[9] *Regulatory Impact Analysis: Control of Emissions of Air Pollution from Highway Heavy-Duty Engines.* EPA420-R-00-010. July 2000. Available online at: http://www.epa.gov/otaq/hd-hwy.htm#regs.

[10] *Final Regulatory Analysis: Control of Emissions from Nonroad Diesel Engines.* EPA420-R-04-007, May 2004

[11] Union of Concerned Scientists, 2004. "Sick of Soot: Reducing the Health Impacts of Diesel Pollution in California." Cambridge, MA. Available online at www.ucsusa.org

Figure II.1.
Contribution of Mobile Sources to NO$_x$, VOC, and PM Emissions

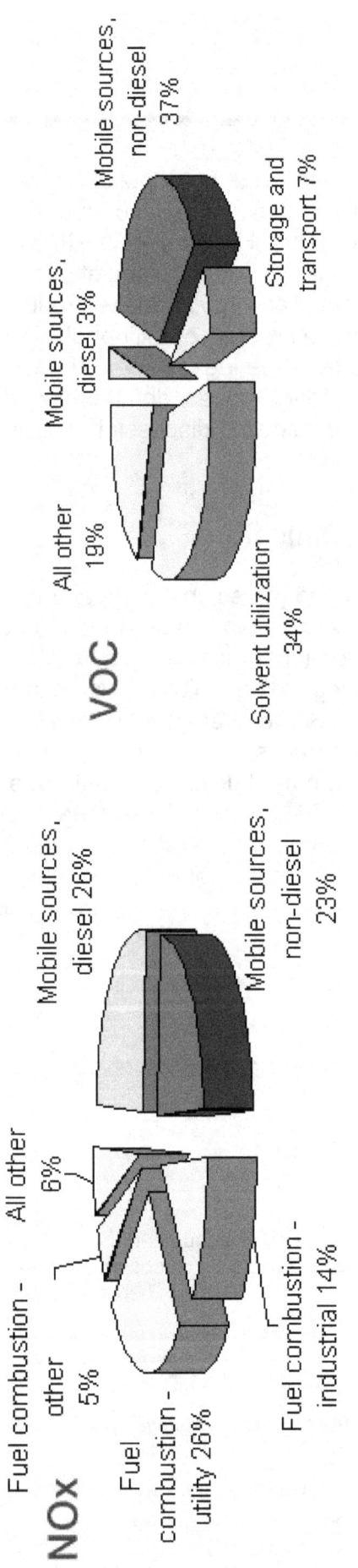

NOx

Fuel combustion - other 5%

All other 6%

Fuel combustion - utility 26%

Fuel combustion - industrial 14%

Mobile sources, diesel 26%

Mobile sources, non-diesel 23%

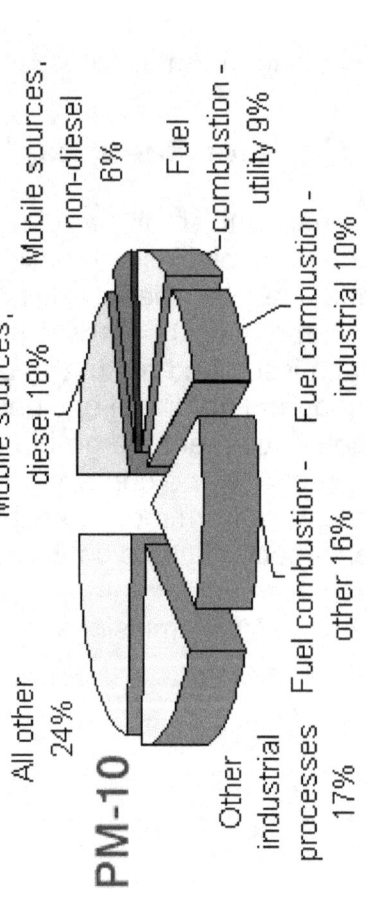

VOC

All other 19%

Mobile sources, non-diesel 37%

Mobile sources, diesel 3%

Storage and transport 7%

Solvent utilization 34%

PM-10

All other 24%

Other industrial processes 17%

Fuel combustion - other 16%

Mobile sources, diesel 18%

Mobile sources, non-diesel 6%

Fuel combustion - utility 9%

Fuel combustion - industrial 10%

Figure II.2

2004 Mobile Source Diesel by Sector

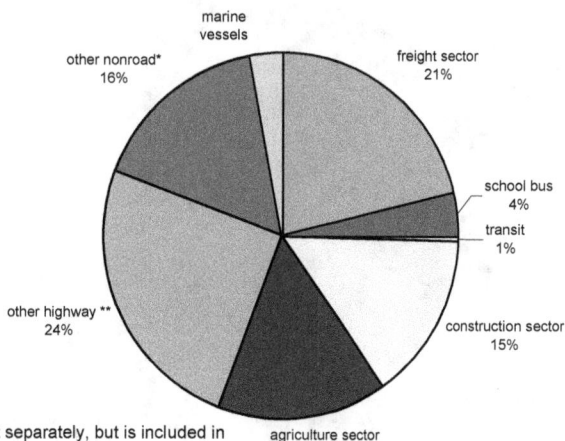

Note: Port data cannot be broken out separately, but is included in relevant sectors
*examples of nonraod include nonroad equipment used at industrial sites and airports
**other highway is smaller trucks and vehicles (LD to Class 5)

Figure II.3

2004 PM2.5 Emissions by Mobile Diesel Sectors

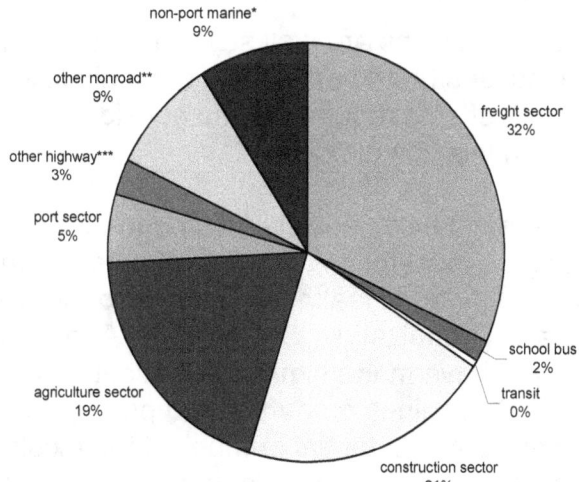

* non-port marine includes recreational vessels and a fraction of C1, C2 and C3 marine
**examples of other nonroad include equipment used at industrial sites and airports
*** other highway refers to smaller trucks and vehicles (LD through Class 5)

5

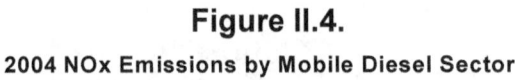

Figure II.4.

2004 NOx Emissions by Mobile Diesel Sector

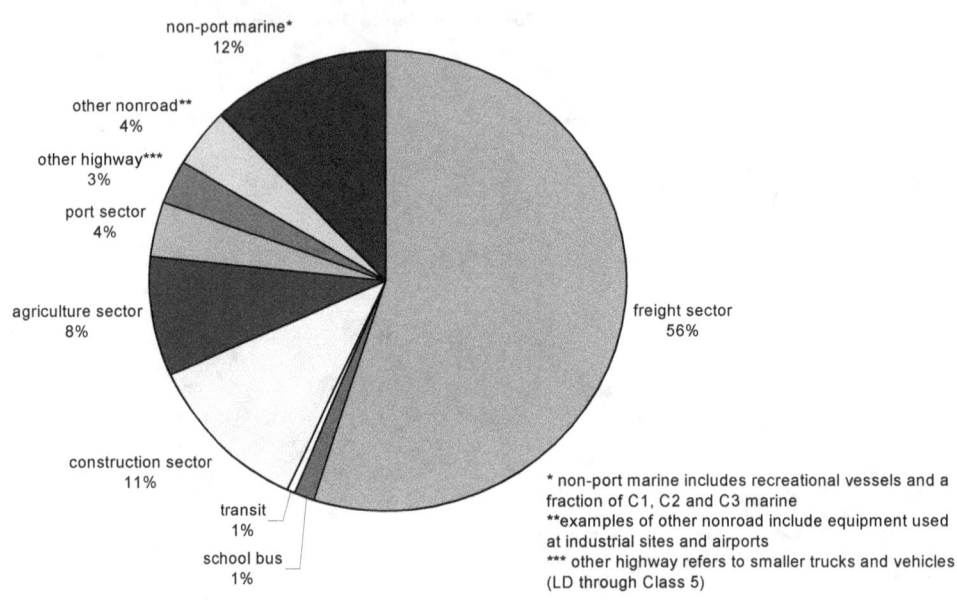

* non-port marine includes recreational vessels and a fraction of C1, C2 and C3 marine
**examples of other nonroad include equipment used at industrial sites and airports
*** other highway refers to smaller trucks and vehicles (LD through Class 5)

1. Health Considerations

The health effects of diesel emissions are well studied, but complex. The level and duration of exposure that causes harm varies from one substance to the next. Precise comments on health effects require careful consideration and the reader is encouraged to read more on this complex issue.[12]

EPA has designated diesel exhaust as a likely carcinogen to humans by inhalation at environmentally adequate exposures. A number of other agencies (National Institute for Occupational Safety and Health, the International Agency for Research on Cancer, the World Health Organization, California EPA, and US Department of Health and Human Services) have made similar classifications. EPA believes its conclusions apply generally to engines manufactured prior to the mid-1990s. As cleaner diesel engines replace a substantial number of the existing engines, the general applicability of the conclusions in EPA's health assessment documents will need to be re-examined. These assessments are periodically reviewed as new scientific studies become available.

---

[12] *US EPA Diesel Hazard Assessment Document for Diesel Engine Exhaust.* 2002. EPA600-9-90-057F Office of Research and Development, Washington< DC. This document is available electronically at http://cfpub.epa.gov/ncea/cfm/recordisplay.cfm?deid=29060

The following sections further describe the potential impacts of diesel exhaust components, specifically particulate matter (PM), ozone, air toxics, and carbon monoxide.

Particulate Matter. PM is another name for particles found in the air, including soot and liquid droplets. Some PM can be large enough to be seen, while others are so small that individually, they can only be detected with sophisticated analytical tools. Particles can be emitted directly from diesel engines (i.e., primary PM) or formed in the atmosphere from gases such as sulfur dioxide (SO_2) or NO_X emitted from diesel equipment (i.e., secondary PM).

Scientific studies have linked certain exposures to PM to various health problems, including aggravated asthma, decreased lung function, increased respiratory problems like chronic bronchitis, and even premature death. Diesel exhaust PM is of specific concern because it has been judged to pose a potential lung cancer hazard for humans as well as a hazard from respiratory effects such as pulmonary inflammation.[12]

Ozone. Ground level ozone (smog) is typically not emitted directly into the air but formed by a chemical reaction between NO_X and volatile organic compounds (VOCs) in the presence of heat and sunlight. NO_X and VOCs are both precursors to smog. Nitrogen oxides are also significant contributors to acid deposition, eutrophication of coastal bodies of water, fine particulate emissions, and haze.

EPA's assessment of scientific studies indicates that ozone can irritate lung airways and cause inflammation, wheezing, coughing, or breathing difficulties during outdoor activities. Repeated exposure to ozone over time may cause permanent lung damage. Even at very low levels, ground-level ozone triggers a variety of health problems including aggravated asthma, reduced lung capacity, and increased susceptibility to respiratory illness. Ozone exposures have been linked to increased hospitalizations and emergency room visits for asthma attacks and mortality.

Air Toxics. The Clean Air Act has no requirements for National Ambient Air Quality Standards (NAAQS) for air toxics (see discussion in following section), but toxic air pollutants can be emitted from diesel engines, as well as alternatively-fueled engines, and are known or suspected to cause cancer or other serious health effects. Examples of air toxics include diesel PM, benzene, 1,3-butadiene, acetaldehyde, POM, and trace metals such as cadmium and chromium.

Studies show that people exposed to toxic air pollutants at sufficient concentrations and durations may have an increased chance of experiencing serious health effects, including cancer. Other health effects can include damage to the immune, neurological, reproductive, developmental, and respiratory systems.

Carbon Monoxide. Once inhaled, carbon monoxide binds to hemoglobin, the substance in blood that carries oxygen to cells. It reduces the amount of oxygen reaching the body's organs and tissues. Exposure to high levels of carbon monoxide can affect mental alertness and vision. People with cardiovascular disease experience chest pain and other cardiovascular symptoms.

2. Environmental Considerations

NAAQS. The Clean Air Act requires EPA to set NAAQS for pollutants considered harmful to public health and the environment. PM, ozone, SO_2, CO, and NO_x have national standards that are set to protect public health with an adequate margin of safety. Areas where air pollution persistently exceeds the NAAQS may be designated nonattainment areas. states with nonattainment areas must develop state implementation plans (SIPs) to ensure emissions are reduced to meet the NAAQS. State and local areas that are responsible for former nonattainment areas, known as maintenance areas, must also develop and implement plans to assure that the areas will continue to comply with the NAAQS. This is especially important in regions with increased population and industrial growth.

Figure II.5. Ozone and PM Nonattainment Areas

On April 15, 2004, EPA designated 474 counties, home to 159 million Americans, nonattainment with the health-based 8-hour ozone standard. [13] On June 29, 2004, EPA also preliminarily found some 244 counties representing 99 million Americans out of compliance with the health-based particulate

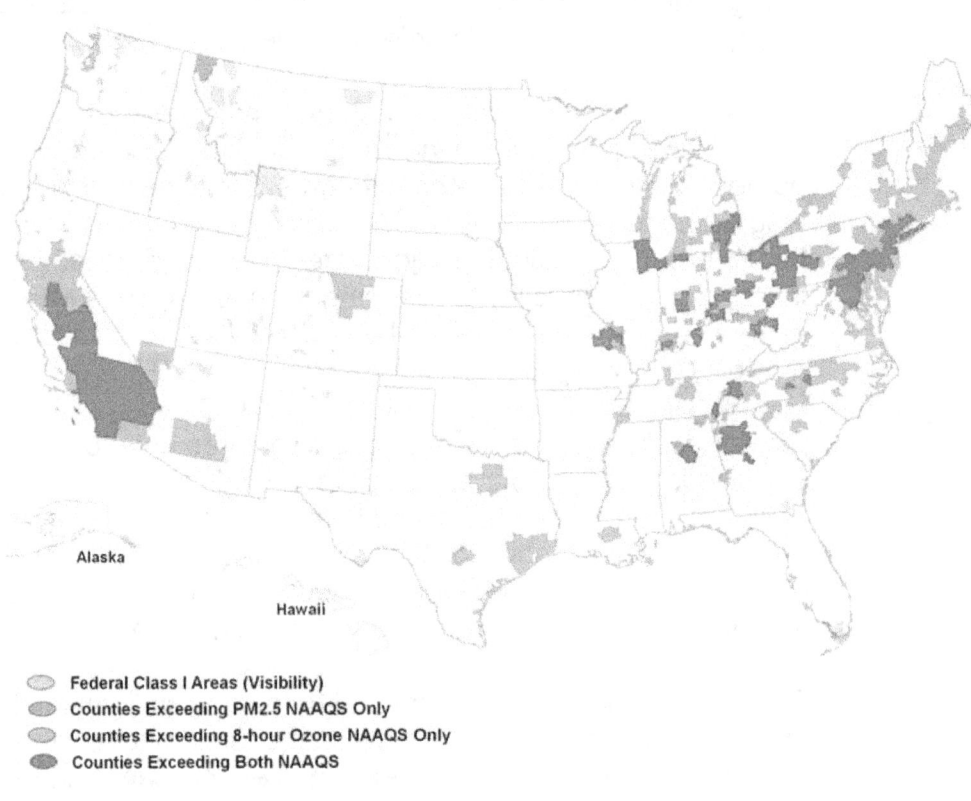

_____

[13] www.epa.gov/ozonedesignations

8

matter standard (see Figure II.5). [14] For the states and local communities that are struggling to eliminate tons of pollution to meet Federal health-based air quality standards, reducing pollution from existing diesel vehicles and equipment is very important. Strategies to voluntarily reduce diesel pollution are a cost-effective way to ensure healthy air.

NATA. Air toxic information (including diesel PM) has been estimated through a national scale assessment known as the National Air Toxics Assessment (NATA). Information is available at www.epa.gov/ttn/atw/nata/.

AQI. EPA calculates an "Air Quality Index" (AQI), which provides information about pollution and public health for five pollutants at the community level. The AQI values can range from 0 to 500—the higher the value, the greater the concentration of air pollution and the greater the health concern. The EPA has developed a website (AIRNow: www.epa.gov/airnow) to provide the public with easy access to national air quality information, both real-time measured conditions and forecasted conditions, which includes AQI information for the current and next day.

---

[14] www.epa.gov/pmdesignations The PM nonattainment areas became final in December 2004.

B. Description of Diesel Emission Reduction Technologies and Strategies

Many technologies and fuels are available for reducing diesel emissions. Some technologies are primarily used to reduce PM while others specifically reduce NO_X. The key is to know the capability of the technology and how well it will work on a given engine to produce the desired results. Proper engine maintenance is always important to ensure appropriate performance of all technologies, for example engine with high oil consumption rates should be repaired prior to installing retrofit technologies.

The following sections describe various technologies available to reduce emissions from existing engines. Appendix B provides detailed comparison of technologies available for ports and construction.

1. Technologies

Diesel Oxidation Catalysts (DOCs) are the most commonly used exhaust aftertreatment technology. A DOC is a catalyzed flow-through metallic or ceramic substrate. A DOC uses catalytic reactions to convert pollutants to water and carbon dioxide (CO_2). A DOC can reduce PM by 20-50% and HC and CO by up to 90%.

A number of DOCs are already verified by the EPA and CARB. DOCs are often selected because they may be used with a variety of fuels, but they generally achieve greater levels of reduction with lower sulfur fuels. DOCs may be used in most applications, and installation is relatively straightforward with very little maintenance required. DOCs perform well on equipment with variable duty cycles.

Diesel Particulate Filter (DPF) is a device that collects and burns exhaust PM at high temperatures. Prior to installation of a DPF, data logging must be performed to ensure the exhaust temperature of the vehicle meets the appropriate specifications. Monitors are required to track exhaust back pressure and exhaust temperature. DPFs generally require periodic cleaning of accumulated ash, which mostly comes from the lube oil and requires special handling. If lube oil consumption is high, more frequent cleaning of the filter will be needed. A high efficiency DPF is desirable because it can achieve a 90% or greater reduction in PM, HC, and CO.

A number of passive and active DPF systems have been verified under the EPA and CARB verification programs. Passive DPF systems continuously or periodically regenerate using the natural exhaust conditions coming from the engine, while active DPF systems utilize heat from another source to burn collected PM. Some passive DPF systems require ultra-low sulfur diesel (ULSD), but all passive systems perform better with cleaner fuels (i.e., the range of passive regeneration is extended when cleaner fuels are used). Active DPF systems utilize fuel oxidation or electrical heating to heat the collected soot to combustion temperatures. The range of some systems has been extended to include both older and newer vehicles.

Partial filters are devices in between a DOC and DPF in terms of PM control, in that they are capable of achieving PM reductions of about 30 to 70%. Filtering is achieved with sintered metal sheets, wire meshes, or in some cases, metallic or ceramic foams. It is recommended that cleaner burning fuels, such as ULSD or lower sulfur fuel be used with DPF's to avoid premature plugging. Although they do not have the same level of PM control as "closed end" DPFs, partial filters have a lower risk of plugging and lower back pressure. Filter regeneration is the same as for high-efficiency DPFs. Similar to DPFs, monitors are required to tract exhaust back pressure and exhaust temperature.

Lean NO_X Catalysts (LNC) systems typically inject diesel fuel (the reductant) into the exhaust stream. The mixture reacts over a catalyst to reduce NOx emissions. LNCs are designed to function effectively at the lean operating conditions found in diesel engines. A LNC combined with a DPF has been verified by CARB. LNC is a relatively new technology and experience is limited. They are reported to have demonstrated NO_X reduction from 10% to over 25% depending on the vehicle operation. If utilized with a DPF, monitors are required to tract exhaust back pressure and exhaust temperature.

Exhaust gas recirculation (EGR) is a technology that can reduce NO_X emissions by up to 50%. An EGR system recirculates an engine's exhaust back to the engine cylinders, which lowers peak combustion temperatures, thereby limiting the production of NOx. Retrofit EGR systems are typically used in conjunction with a DPF to control the resultant higher PM emissions. One low pressure EGR system that incorporates a DPF is currently verified by CARB. Maintenance on an EGR system may be minimal, but DPF-equipped systems still require regular maintenance.

Selective catalytic reduction (SCR) is another technology designed to reduce NO_X emissions. SCR systems inject a reductant (typically urea or ammonia) into the exhaust to facilitate a catalytic reaction with the NO_X on an SCR catalyst. SCR can reduce NO_X emissions by 80%, but appropriate exhaust temperatures and engine operating modes are critical for optimal NO_X reductions. SCR may also be used in conjunction with a DPF to reduce PM. An SCR system has been verified under CARB's program for a select number of engines. A monitor/controller will be required to control injection of the reductant and monitor back pressure and temperature.

Crankcase emission control technologies can be retrofitted on engines to eliminate crankcase vent (CCV) emissions. Historically, turbocharged diesel engines have vented crankcase emissions to the engine compartment and below the vehicle. Crankcase emission control technologies may filter exhaust from the crankcase and re-route the filtered air back to the intake, thereby reducing crankcase PM. In fact, total (i.e., exhaust and crankcase) PM emissions may be reduced by 5-10% or more. Both open and closed systems are available on the market. There may be maintenance associated with some of these systems. One CCV system has been verified by the EPA and CARB verification program in combination with a DOC.

CCV are especially important on older school buses, even those retrofitted with DOCs or DPFs. The University of Washington particulate research center found higher levels of engine emissions inside school buses, especially when the windows were open.[15]

2. Fuels

ULSD contains less than 15 parts per million (ppm, by weight) sulfur. It enables catalyst-based and other emission reduction technologies to operate at maximum effectiveness. Even without the use of an aftertreatment technology, ULSD can reduce PM emissions by 5-10% compared to standard on- or nonroad diesel fuel. Beginning late in the summer of 2006, all on-road diesel fuel will be phasing down to ULSD from the current 500 ppm, or low sulfur, diesel (LSD). Nonroad diesel fuel standards will be gradually phased in to lower the sulfur content until 2010, at which time most diesel fuel will be ULSD.

Biodiesel is a domestic renewable distillate fuel derived from a number of vegetable oils, animal fats, or used frying oils. Biodiesel is typically blended with petroleum-based diesel fuel, usually with blends ranging up to 20% biodiesel, referred to as B20. Since the biodiesel base stock can vary, the specific fuel properties vary depending on the biodiesel source and the degree of processing refinement. Typically, B20 provides about a 10-15% reduction in PM and a 0-10% reduction for CO and HC. However, in testing emissions from heavy-duty engines using biodiesel fuel, EPA found that NO_X emissions can increase depending on the type of base stock and portion of biodiesel. Some more recent studies have indicated that using biodiesel fuel can either show an increase or no effect in NOx emissions, but the factors affecting NO_X emissions levels have not been clearly determined. Biodiesel was generally verified by EPA and the level of PM, HC, and CO reduction is related to the portion of biodiesel used.

One of the current issues regarding biodiesel and other alternative fuels is uncertainty of their effects on engines and emissions for the new advanced engine and aftertreatment systems required by EPA regulations starting in 2007. The impact of using biodiesel blends in ULSD burned by these new engine systems needs further investigation.

Emulsified diesel fuel is a blended mixture of diesel fuel, water and other additives. It can be used in most diesel engine applications, but some reduction in power and fuel economy is expected due to the fact that the addition of water reduces the energy content of the fuel. Emulsified diesel can reduce NO_X and PM emissions by about 20- 50%, especially when used synergistically with aftertreatment. Engine

---

[15] Hill LB, Zimmerman NJ, and Cooch J. A Multi-City Investigation of the Effectiveness of Retrofit Emissions Controls in Reducing Exposures to Particulate Matter in School Buses. Clean Air Task Force report, January 2005. Jackson, NH.

manufacturers may have requirements for usage of this fuel, and should be consulted prior to use. Emulsified diesel fuel products have been verified for use by EPA and CARB.

CNG is an alternative fuel consisting mostly of methane and odorless and colorless. CNG, requires a special infrastructure, and is available at approximately 1,300 refueling stations. Emissions reductions can range from 35-60% for NO_X emissions and 70-90% for PM emissions.

Liquefied Natural Gas (LNG) is similar to CNG in that it too is odorless, colorless, and composed of mostly methane. Most LNG vehicles are used by fleet managers, thus refueling infrastructures are located at the fleet operation site and not available to the general public. LNG can reduce NO_X emissions by approximately 50%.

Propane or Liquefied Petroleum Gas (LPG) is a byproduct of natural gas processing and petroleum refining. It burns more cleanly than gasoline, but its supply is limited. Propane-fueled vehicles are already common in many parts of the world.

Idle reduction technologies can be very effective strategies to reduce emissions including greenhouse gases. These operational strategies can reduce wait and loading times for cargo and passenger vehicles. Add-on devices are available that reduce idling on long haul trucks, as well as fixed equipment that provides electricity to heat and cool trucks and their loads.

C. Description of Incentives and Strategies Considered

EPA has set a goal of achieving maximum reductions from the legacy fleet over the next 10 years. The Work Group agrees cleaning up pollution from these 11 million engines will require substantial investment in the range of $50–100 billion. However, the Work Group believes the task is not insurmountable and is one worth doing in terms of return on investment to society. Each year, owners and operators of the legacy fleet spend over $100 billion to operate and maintain existing engines and vehicles. For just a fraction of what is spent, perhaps as little as 5%, substantial gains could be achieved in reducing emissions from existing engines and vehicles.

A variety of incentives are available for reducing diesel emissions, but none of them provide a "silver bullet solution" that will reach every machine, vehicle, or truck, or please every stakeholder involved. However, by combining incentives and tailoring them to specific sectors, many of the incentives outlined below can or do work to reduce emissions.

Table II.1 presents a summary of incentives that are or have the potential to be available for each sector addressed in this report.

Table II.1. Summary of Incentives and potential to apply in the sectors

Type of Incentive	Industry Sector			
	School Bus	Construction	Ports	Freight
Tax Related Incentives		X	X*	
Government Grants and Rebates	X	X	X	X
Supplemental Environmental Programs	X (State)	X	X	
Publicly Funded Cleaner Fuels		X	X	
Voluntary Contract Modifications		X		
Low Interest Loan Programs		X	X**	X
Contract Requirements	X	X	X	X
Regulatory Credits		X	X	
Public Recognition, Environmental Stewardship and Non-Monetary Incentives	X	X	X	X
Regulatory and Mandatory Requirements	X	X	X	X

* May be applicable to marine port terminal operators which are private entities
**May be applicable to trucking operations at marine ports.

Following are summaries of incentives under consideration. With some notable exceptions, most diesel emission reducing activities require a financial investment. For this reason, incentives are broken into three categories based on what type of entity bears the majority of the economic cost: primarily government-funded Incentives, government- and private sector-funded incentives, and primarily private sector-funded incentives. Regulatory and mandatory requirements, current regulatory programs, and other strategies are also described.

1. Primarily Government Funded Incentives

Income Tax-related Incentives. Tax incentives help offset the cost of reducing diesel emissions by reducing the amount of taxes a taxable entity would pay. Tax incentives can take the form of tax exemptions, tax deductions (including accelerated depreciation), or tax credits. Tax exemptions exclude certain items or activities from being taxed, while tax deductions and accelerated depreciation reduce the taxable income for certain expenses. Tax credits directly reduce tax liability based on the amount of expense.

Typically such measures set no deadlines and require no applications, providing time for manufacturers to respond and flexibility for the interested owners of equipment. Tax incentives offer relative ease of use to profitable or taxable entities that qualify for the incentives.

Significant government functions are needed to establish and maintain tax incentives. In addition, tax incentives can be more difficult than other measures to target to specific applications or geographic areas where they may be most needed. The incentive must also be large enough to motivate qualifying entities to take advantage of it. . Since efficiency gains are generally not realized from the retrofit of diesel equipment or use of alternative fuels and, therefore, no return on investment, companies might not be motivated by a tax incentive of less than 100%. However, a tax incentive of less than 100% could be successful if applied for a fleet modernization strategy Tax incentives at the state level (e.g., Oregon and Georgia) have been largely unable to garner participation due the small amount of financial incentive.

Excise Tax-related Incentives. Governments impose other taxes that can be reduced or eliminated to encourage the use of less polluting technologies or fuels. For example, the recently passed Federal transportation legislation, referred to as SAFETEA-LU (Pub L 109-59), includes a 50 cent-per-gallon (or gasoline gallon equivalents (GGE) in the case of compressed natural gas (CNG)) excise tax credit for every gallon or GGE of non-petroleum alternative fuel used. This excise tax credit is taken by the fuel seller. However, in those cases where the seller and user are the same (such as when a school district owns and operates its own fueling station) the excise tax credit goes to the user. The legislation also provides that the amount of the credit shall be paid to the entity entitled to the credit; it is remitted to the seller as a quarterly check.

Government Grants and Rebates. Grant programs provide funding directly to equipment owners to allow them to reduce diesel emissions in their fleet. Rebates are a type of grant in which a governmental or nonprofit entity establishes reimbursement specifications for projects that could reduce emissions. Typically, the government or nonprofit entity announces the availability of a predetermined number of rebates at a set funding amount. Operating and maintenance costs have not typically been covered by grants or rebates.

Grant programs can be highly effective in achieving targeted, cost-effective emissions reductions and can leverage matching funds, thereby creating a partnership for sharing the responsibility of reducing emissions. However, grant programs can be difficult to start up and resource intensive to implement and administer to ensure the emissions reduction. It is difficult to provide funding directly to the private sector at the federal level, so federal funds would most likely help retrofit government fleets or be passed through a state or local agency or nonprofit organization. Utilizing rebates may help alleviate some of the administrative burden of grants for both governments and grant applicants.

In a rebate system, a governmental or non profit entity establishes rebate specifications and announces the availability of a pre-determined number of rebates at a set funding amount for particular types of projects that reduce emissions. For example, State Q may provide up to $1,000 each to the first 500 applicants who will implement strategy X, Y, or Z. Rebate programs need to be structured carefully in order to ensure that the financial benefit ultimately flows to the technology user, and that overall economic development is not discouraged. Nonprofit co-ops could be utilized to help small businesses apply for clean diesel grants. Both grants and rebate programs often suffer from the vagaries of the annual appropriations process unless dedicated funding streams are enacted.

Examples of grant programs include California's Carl Moyer Program, the Texas Emissions Reduction Plan, the Ports of Long Beach and Los Angeles Gateway Cities Clean Air Program, EPA's Clean School Bus USA, the National Clean Diesel Campaign, and Idle Reduction Grant Programs. CARB estimates that the Carl Moyer Program reduced NO_X emissions by about 14 tons per day at a cost of about $3,000 per ton. Though the historical focus of the program has been NO_X, funding for engine/vehicle replacement has also reduced PM by 1 ton per day. These benefits accrue from each project for a minimum of 5 years. As of June 2005, the Texas Emissions Reduction Plan has granted more than $183 million dollars towards diesel reduction projects that average roughly $4,600 for every ton of NOx reduced.

Supplemental Environmental Projects in Settlements of Legal Actions against Environmental Violators. A Supplemental Environmental Project (SEP) is a project that is negotiated as part of a legal settlement in litigation against environmental violators. In order for a project to be eligible for inclusion as a SEP, it must have nexus to the violation that has occurred and must be administered by the defendant in the litigation.

SEPs have been used to reduce emissions from school buses and other types of diesel engines. They can be quite large and achieve important reductions in diesel emissions. For example, the federal government and Toyota agreed to a $20 million SEP for school bus retrofits. Some states, including Illinois, Massachusetts, and Connecticut, have also successfully included SEPs in environmental settlement agreements.

Congestion Mitigation Air Quality (CMAQ) Funded Projects. CMAQ is a set-aside under the Surface Transportation Program in the Highway Trust Fund, which is funded from the fuel tax. Its express purpose is to reduce pollution and congestion in areas that are designated as NAAQS nonattainment or maintenance. CMAQ money is apportioned by a formula set by Congress and is used by metropolitan planning organizations (MPO) to fund a variety of projects in their geographic area, including retrofits. The MPO selects the projects to be funded. For the first time, the most recent SAFETEA-LU (Pub L 109-59) specified that CMAQ money may be used for reducing pollution from nonroad equipment used in construction projects funded from the Highway Trust Fund. CMAQ is currently authorized at over $8.4 billion for a 6-year period beginning in FY2006.

Publicly Funded Cleaner Fuels. Instead of contractors bearing the cost of cleaner fuel, a contracting entity could provide cleaner fuel at the cost of the less clean fuel. This incentive shifts the financial burden of purchasing cleaner fuel onto the entity requesting services, such as a municipality. The provider of the cleaner fuel could subsidize the incremental cost above what the contractor/operator would normally spend on diesel. However, municipalities have very limited resources to subsidize and distribute fuel, especially for very large operations.

2. Government and Private Sector Funded Incentives

Voluntary Contract Incentives, Bonuses and Allowances. Voluntary contract incentives provide a mechanism for state and local governments to reduce diesel emissions from public works projects by offering a bonus or providing an allowance to contractors who are willing to retrofit their fleets. Contract incentives, bonuses or allowances are distinguished in this section from contract or lease requirements or other mandatory contractual practices. Contract allowances incorporate a payment to the contractor to offset, fully or partially, the cost of emission-reducing activities. It should be noted that the financial burden of reducing emissions could be placed either on the governmental contracting entity or the private sector depending on the design of the contract modification.

The contracting community views voluntary contract incentives as being more accommodating to small business concerns. Although small businesses prefer voluntary provisions rather than mandates, even voluntary provisions can result in competitive disadvantage for small businesses with limited resources. This is especially a problem for public entities that are required to provide a fair share of their business opportunities to small and minority-owned businesses.

Low Interest Loan Programs. Low interest loans could help provide the necessary capital for emission-reducing activities while minimizing the long-term financial burden of a financial assistance program. They could be administered through a governmental entity, port authority, or in a public-private partnership with a bank. In a revolving loan program, the net interest paid over time could be used to fund other projects.

Loan programs may not be an attractive incentive for retrofit projects that do not have a direct or indirect positive economic impact on the borrower unless another motivating factor is provided for reducing emissions (such as contract modifications, mandatory requirements, etc.). However, a loan program may be appropriate for emission-reducing activities that have an economic benefit such as fuel savings. Low interest loan programs could also be particularly useful for small businesses in providing capital. Low-interest loans have the greatest impact if coupled with other incentives like grant programs

3. Primarily Private Sector Funded Incentives

Regulatory Credits. Regulatory credits provide some kind of regulatory relief or flexibility in exchange for reducing emissions, and require cooperation between private and public sector entities. Regulatory credits include State Implementation Plan (SIP) credits, conformity credits, Mobile Source Emissions Reduction Credits (MERCs), and Supplemental Environmental Projects (SEPs). SIP credits are emissions reductions that are counted toward a state or locality's required emissions reductions for meeting Federal air quality standards, and conformity credits are emissions reductions required for projects that would otherwise result in an overall increase in emissions.

Governmental entities and public port authorities can be motivated by SIP and conformity credits to reduce emissions. Interest exists among public entities to get credit for early voluntary action. Private entities, on the other hand, would be more likely to utilize the tradable permit system of MERCs or conduct a SEP in lieu of paying the full cost of an environmental enforcement action.

The challenge for utilizing MERC, SIP, and conformity credit is the requirement that the emissions reductions be quantifiable. In this regard, public port authorities and others have requested guidance and recognition for claiming credits. However, credit trading programs raise concerns regarding the inability to ensure emissions reductions in a particular location, as well as accountability issues related to the use and mobility of equipment.

Public Recognition, Environmental Stewardship and Non-Monetary Incentives. Non-monetary incentives like public recognition can also be attractive to some fleet owners/operators for a host of reasons. Government agencies often encourage non-monetary incentives by providing public recognition as well as educational information and technical assistance.

Positive emission-reducing actions, however, do not need to simply be altruistic. Operational efficiencies that reduce emissions often make good business sense. Examples include adopting an Environmental Management System (EMS) that provides a framework to integrate environmental decision making into an organization's operations. In addition to taking a multi-media approach to mitigating environmental effects, an EMS can often result in long-term cost savings.

4. Regulatory and Mandatory Requirements

Mandatory requirements can take several forms, the most familiar of which is a federal or state regulation setting new engine emission standards or requiring after-treatment technology. Regulatory requirements provide the opportunity to target specific areas. Like incentives, they can also impact private fleets. Significant government functions are needed to establish and maintain such requirements. A good regulatory process allows all impacted parties, including industry, public health and environmental groups, and members of the public the opportunity to provide input into the development of the regulations. The regulatory process can promote overall economic efficiency by comparing the costs of compliance with the public health benefits.

All Work Group members acknowledge that regulatory mandates are one approach to achieving air quality benefits. However, they disagree about who should pay for the costs of retrofits required by regulation. Some members believe that the end users should pay for the retrofits and that this principle is well-grounded in the tradition of regulatory mandates. Others believe that, for regulatory approaches like contract specifications, governments should provide funding mechanisms to support the implementation of the specifications. Still others believe that it is unreasonable to require end users to invest in retrofit equipment for engines that met all of the regulatory requirements at the time of original purchase, regardless of the funding issue.

Having noted this difference of opinion, the Work Group agrees that these philosophical differences are better addressed in the political process. It should also be noted that the EPA's authority to regulate the legacy fleet differs significantly from its authority to regulate new engines.

Regulation of Highway Vehicles. At the federal level, EPA has the authority to set emission standards for both on- and nonroad new engines and vehicles. However, questions do exist regarding EPA's authority to regulate the in-use fleet for highway engines and vehicles. Section 202(a)(3)(D) of the Clean Air Act (CAA) gives EPA authority to set requirements for engines at the time of engine rebuild, but regulatory authority to implement retrofits more broadly needs further review.

Under the CAA, only California may set its own emission standards for new highway engines, subject to receiving a preemption waiver from EPA under Section 209(b). Other states may adopt California standards pursuant to the terms of section 177 of the CAA. States generally can adopt provisions relating to the use, operations, or movement of engines and vehicles within their borders such as carpool lanes.

In the court case <u>Allway Taxi Inc. v. New York,</u>[16] the Federal District Court held that a state or locality can not impose its own emission control standards the moment after a new car is bought and registered, as that would constitute an obvious circumvention of the CAA and Congressional intent to prevent obstruction of

---

[16] Allway Taxi Inc. v. New York, 340F.Supp. 1120 (S.D.N.Y.), add'd 468 F.2d 624 (d2. Cir.1972)

interstate commerce. The District Court stated that the sections preempting states form setting standards for new vehicles do not preclude a state or locality from imposing its own exhaust standards upon the resale or re-registration of the vehicle.

In related recent rulings, the U.S. Supreme Court in <u>Engine Manufacturers Association v. South Coast Air Quality Management District</u>[17] held that requirements mandating a private operator's purchase of alternative-fueled vehicles constitute a type of emissions standard that states and political subdivisions are preempted from adopting under Section 209(a) of the CAA. On remand, the U.S. District Court stated in its order denying a motion to implement the Supreme Court decisions that purchase requirements as applied to state and local government fall within the market participation exemption to preemption and are not preempted by Section 209(a).

Regulation of Nonroad Vehicles and Engines. For nonroad vehicles and engines, EPA can set new engine standards under CAA Section 213, but does not have any statutory authority to set standards for in-use engines. California can regulate certain new and non-new nonroad engines provided that it first obtains authorization to do so under CAA Section 209(e)(2). No state, including California, can regulate new engines used in construction and farm equipment under 175 horsepower (hp), new locomotives, or new engines used in locomotives. In addition, no state other than California may set standards for nonroad spark-ignited engines smaller than 50 hp. Other states may adopt California's new or non-new nonroad standards that have been authorized by EPA with the exception of spark-ignited engines smaller than 50 hp.

All states can control the use, movement, and operation of registered nonroad vehicles within their borders with the exception of locomotives. Locomotives present unique challenges and are not addressed in this document. California may request authorization (i.e., apply for a waiver) under Section 209 (e)(2) to establish retrofit programs for in-use nonroad engines and vehicles, and other states may adopt California's program.

Federal, state and local regulatory agencies are limited in their authorities to regulate ocean-going vessels, especially vessels flagged in foreign countries. Regulations applicable to ocean-going vessels are established by means of international treaties.

5. Current Regulatory Programs

California Air Resources Board Retrofit Regulatory Program. As part of California's Diesel Risk Reduction Program, CARB has developed and implemented several rules and regulations to control PM from some diesel mobile sources, including waste collection trucks, transit agency vehicles, and portable engines. For example, CARB requires cleaner engines, cleaner fuel and the retrofitting of older

---

[17] Engine Manufacturers Association v. South Coast Air Quality Management District, 124 S. Ct 1756 (2004)

buses in transit fleets. Waste collection haulers are given a choice of several options for meeting the "best available control technology" standards. School buses are subject to idling restrictions for new and used engines. CARB continues to expand these mandates to include more applications.

CARB is currently in the planning and development stages for devising rules and regulations on several in-use diesel sources, including nonroad and cargo handling equipment, on-road trucks, and some marine applications. In-use requirements for cargo handling equipment and heavy duty vehicle idling restrictions are expected to be adopted in late 2006 and implemented in early 2007. Clean fuel requirements for ocean-going vessel auxiliary engines are also expected to be approved late 2005 or early 2006 and implemented in late 2006 or early 2007. For in-use nonroad equipment measures, CARB is currently conducting surveys of equipment, performing field research, and discussing regulatory concepts with the regulated community.

6. Other Strategies

Contracting Requirements. Both state and Federal governments have stipulated required diesel emission reduction activities as a part of a contract's terms and conditions. Contract preferences establish bid evaluation criteria that favor cleaner contractors. While these contractual performance requirements would help guarantee emissions reductions, business groups are often concerned that these requirements hamper the ability of small businesses to compete because many do not have the necessary resources to meet the requirements. This concern can be at least partially mitigated if adequate funding is made available to the small business. Similarly, contract or lease requirements between a landlord port and their tenants could require emission-reducing activities as part of the business agreement. Seaport terminal leases are often established for as long as 30 years, and offer limited and inequitable opportunities as tools to reduce emissions.

The Clean Diesel and Retrofit Work Group discussed but did not reach consensus on regulatory and mandatory contractual requirements for emissions reduction activities. Some members expressed the opinion that incentives cannot, standing alone, achieve the desired reductions in pollution from the legacy fleet. Other members took the position that it would be premature to reach that conclusion and that the boundaries of EPA regulatory authority should limit consideration of Federal regulatory strategies in this report.

Other Tax and Fee Strategies. Governments can influence decisions on purchasing clean vehicles as well as cleaning up existing engines through a combination of fees and similar strategies. For example, in Europe, a road tax is higher for older vehicles. In California, as well as other countries, registration fees are higher for higher-polluting vehicles. Fuel taxes can also be used and rebated to generate a revenue stream for cleaning up existing engines.

III. Summary of Key Sector Recommendations and Cross Sector Incentives

Concurrent with the work of the Clean Diesel and Retrofit Work Group, Congress has passed the Energy Bill (Pub L 109-49) and the SAFETEA-LU (Pub L 109-59), both of which recognize the importance of reducing diesel emissions from the legacy fleet as well as the need for more funding. Several of the recommendations of this Work Group, specifically grants and loans for retrofit and replacement, have been authorized by Congress to be funded at levels in excess of $200 million per year. As discussed previously, the SAFETEA-LU includes provisions that make Congestion Mitigation and Air Quality (CMAQ) Funding ($8.4 Billion over 6 years) available for reducing emissions from diesel engines and vehicles used in Construction projects built with funds authorized under the highway trust fund. Since the context in which these recommendations were formulated has changed significantly, the Work Group is considering the impact of these bills on its recommendations. However, the following summarizes the recommendations to date.

A. General

- The potential benefits of cleaning up the legacy fleet are significant and worth large scale public investment.
    - Public funds should be used to creatively leverage other investments.
    - The Work Group would like to see retrofit programs fully resourced, including staff to run the programs.

- Given the breadth of applications and uses of diesel engines, and the mix between public and private fleet owners across the various sectors examined, it is important to provide a range of options for addressing diesel emissions to each sector.

- Maximize emissions reductions in each situation given the air quality needs and technical feasibility.

- The SAFETEA-LU (Pub L 109-59) and Energy Bill (Pub L 109-49) provide new opportunities for addressing diesel emissions from all sectors, and the members would like to explore these opportunities and assist states and localities to take full advantage of them.

B. Cross Sector Recommendations

- All of the sector sub-groups have identified *Grants, Loans and Rebates* as attractive incentives. The Work Group is committed to advocate for establishing such programs and is willing to participate in designing effective programs at all levels.

- *Tax Incentives* were identified as having broad appeal to private fleet owners and operators. Tax incentives can bolster the business case for retrofits and replacement, and reduce the inherent risks for cleaning up equipment. They are appropriate to pursue at the Federal level as well as other levels of government.

- *Outreach and Education* was identified by all sectors as key to getting emission reduction strategies in place. Regardless of whether it is a grant, loan, rebate or tax credit, people need to know the benefits of reducing diesel emissions, how to access available resources, and what technology best applies to engines and vehicles in their situations.

- All sectors identified a National Recognition Program as having the potential to promote diesel reductions, especially if that program was designed to ensure positive publicity and prestige.

- *Enhanced Technology Verification.* To ensure that the best technologies are made available as quickly as possible, the national technology verification process can be streamlined to move new technologies into the market. Work Group members are willing to assist EPA in verification process improvements, including working to assess the resource needs to carryout the process.

IV. Sector Analysis and Strategies

A. Clean School Bus Sector Report

The first public school transportation for children began in the late 1800's when local farmers loaned horse-drawn wagons for that purpose. From that humble beginning, the school transportation sector has grown to encompass over 480,000 buses transporting 25 million public school students each day.[18] The first national conference to consider the safety of public school buses was in 1939, when representatives from 48 states gathered to recommend standards. Since that time, the school transportation community, along with the Federal Motor Vehicle Safety Standards that apply to school buses, has made school buses the safest way to get children to and from school and school-related activities.

Of the 480,000 school buses in the nation, approximately 400,000 are large school buses (over 10,000 pound gross vehicle weight rating) that generally are diesel-powered, mostly with regular diesel fuel. About 4,000 of these large school buses are powered by alternative fuels, such as compressed natural gas and propane. Some older, large school buses are powered by gasoline. The others (about 80,000) are small school buses (10,000 pounds GVWR or less), most of which are diesel-powered, but some are gasoline-powered.

As a result of the Clean School Bus USA Program, SEPs, and other diesel retrofit programs; EPA estimates that about 30,000 of the 400,000 diesel-powered large school buses have been involved in a clean school bus project. These buses may be running on a cleaner fuel such as ULSD or biodiesel, may have been retrofitted with emissions control devices, or may have been replaced by an alternative fuel-powered bus.

About one-third of the nation's school buses were built before model year 1991. These buses emit at least six times more PM and twice the NO_X compared to a model year 2005 diesel-powered bus. About 2,000 school buses on the road were built before 1977. These are the Nation's oldest and highest-polluting school buses.

School transportation is provided by the more than 14,000 local school districts in the U.S. Approximately 70% of the school buses in the U.S. are owned, operated and maintained by the school district.[18] The other 30% of the school buses are owned, operated and maintained by private contractors to the school districts.

While individual school districts are responsible for transportation of children to school, in some areas of the country the purchase of parts and/or buses is accomplished through either the state or a group of school districts. A few state

---

[18] The number of school buses in operation, the number of pre-1977 school buses in use, and the split between school buses owned and operated by public school districts versus private contractors are taken from data published in the 2005 issue of *School Bus Fleet Magazine's Fact Book*.

governments purchase new buses for all districts within the state. Some states have blanket purchasing arrangements for buses and/or parts (such as emissions control devices). Boards of Cooperative Educational Support (BOCES) are groups of school districts that collectively purchase materials, such as vehicle parts or fuel. In other areas, this is accomplished through School District Councils.

School transportation is a local responsibility and, thus, funded by local taxpayers as an education-related expense. Since the majority of school districts are already cash-strapped, very few are in the position to be able to afford a clean school bus project. To date, almost all clean school bus projects have been funded by Federal government grant funds, SEPs, or state funds. A few projects funded exclusively by the private sector also have been undertaken.

1. EPA's Clean School Bus USA Program

Diesel exhaust has health implications for everyone. Children are especially sensitive to air pollution because their respiratory systems are still developing and they have a faster breathing rate. Recent studies suggest that children's school bus commutes potentially expose them to significantly higher concentrations of pollutants from various sources (e.g., tailpipe, crankcase, etc.) than what is measured in the community's outdoor air. In addition to tailpipe emissions, some research indicates that the crankcase may be a source of significant on-board exposure, and some exhaust emission control technologies may not have a significant impact on in-vehicle exposure. CARB is currently evaluating the contribution of different sources to in-vehicle exposure, including tailpipe, crankcase and other vehicles on the road. Further studies are necessary.

EPA's Clean School Bus USA program was created in response to concerns regarding children's exposure to diesel emissions from school buses. The Program has three primary goals: (1) reduce school bus idling; (2) retrofit existing buses with devices and/or cleaner fuels that reduce pollution; and (3) replace buses built before model year 1991 with new, cleaner buses, and target first the replacement of school buses built before April 1, 1977. Congress allocated $5 million in both FY2003 and FY2004, and $7.5 million in FY2005, for a cost-shared grant program to upgrade diesel school bus fleets in public school districts. To date, EPA has awarded almost 40 grants to communities across the country for clean school bus projects. EPA anticipates awarding 20-30 more grants in late fall of 2005.

In addition, the program has created public information materials and an informative web site to guide school officials, transportation managers and others in their efforts to establish reduced idling programs and to develop means for retrofitting or replacing diesel-powered school buses in their fleets.

As a direct result of EPA grants under the Clean School Bus USA Program, approximately 10,000 school buses will have been retrofitted, replaced or switched to a cleaner fuel at the end of the grant project period (June 2006). See Table IV.1 below for a breakdown of technology and fuel applications for these grant projects as of December 2004 [note that these numbers are approximate as several 2003 and most 2004 grants are still on-going and subject to change].

**Table IV.1. EPA 2003 and 2004 Clean School Bus Grants:
Technologies and Fuels as of December 2004**

Technology/Fuel	Number of devices/ buses affected
Diesel Oxidation Catalysts (DOCs)	2169
DOCs and Crank Case Ventilation Systems	277
DOCs and ULSD	87
DOCs and biodiesel (any blend)	215
Diesel Particulate Filters (DPFs)	105
DPFs and ULSD	327
CNG Replacements	20
Biodiesel	240
Emulsions	40
ULSD	3969
Total	***7449***

Over 1 million children now ride cleaner school buses, and approximately 20 million residents of communities in which clean school bus projects have taken place are breathing cleaner air. In general, most projects have been straight-forward, with the districts ably navigating both grant requirements and application of the technology and/or fuel. Few technology failures or problems with cleaner fuels have been reported thus far. That said, it is not altogether an "easy" project for school districts to accomplish. Planning, partnerships with other organizations, and dedication to the project help ensure successful implementation.

2. Key Issues

Districts must overcome a number of key issues in order to successfully implement a clean school bus project.

Funding. Upgrading the Nation's diesel-powered large school buses still in need of replacement or retrofit will be very expensive. For example, to replace just one bus with a new clean school bus costs between $75-$100 thousand dollars. School districts simply do not have the funds for a project that is not seen as an absolute necessity – districts will not choose retrofit equipment over teacher salaries or textbooks, nor should they. Therefore, other parties have had to fill the gap. First, Federal funds primarily through EPA, as well as the U.S. Department of Energy, have allowed many communities to implement clean school bus projects. Second, settlements for Clean Air Act violations with companies on both the federal and state level have funded school bus projects across the country. (At present, it appears that SEPs with EPA in 2005 and perhaps beyond are no longer eligible for school bus activity due to issues of possible budget augmentation.) Third, some states, such as California, New York and Washington, have developed funding mechanisms for school bus retrofits and replacements. In addition, the rising cost of petroleum coupled with the $0.50/gallon excise tax credit (which for school districts will operate like a grant program) established by the volumetric Excise Tax Credit for Alternative Fuels in the recently passed Federal SAFETEA-LU (Pub L 109-59) may provide some school districts with a sufficient economic incentive to purchase new alternative fuel school buses.

Knowledge/Skill/Technical Capacity at the Local Level. In order to successfully implement a clean school bus project, personnel within the school district must have some technical capacity with pollution control options and strategies. In addition, personnel must be able to write a successful grant application and handle the additional responsibilities of grant management. Not all school districts, especially those that are smaller and have fewer resources, have the capability to investigate various strategies for retrofitting or replacing diesel-powered school buses in their fleets.

Private Fleets. In significant areas of the country, particularly the Northeast, Mid-Atlantic, upper Midwest, and in large urban centers (most of which are in nonattainment areas), the majority of school buses are owned and operated by private companies under contract to public school districts. Currently, private contractors must apply for Clean School Bus grant funds jointly with a school district. If a school district chooses not to participate in the Clean School Bus USA Program, the private contractor has no way of applying for grant funds, and those communities become ineligible to participate in the program.

Cleaner Fuel Availability and Device Applicability. While mandated to be available nationwide in October of 2006, at present ULSD fuel is available only in areas near refineries or ports from where it can be shipped relatively cheaply, or in areas where the demand is sufficient that fuel suppliers will truck the fuel to fleets. The price per gallon for ULSD compared to regular diesel fuel varies widely, depending on how far the fuel must be shipped and by what mode. For the short term, this limits the use of certain retrofit technologies, namely some DPFs since they must be used in conjunction with ULSD.

DPFs have been verified by EPA and CARB with different temperature specifications, and not all DPFs are appropriate for school bus operations. When applications do not meet minimum temperature specifications, they do not regenerate to burn off the collected PM and may require more frequent maintenance or may fail entirely. In the past, there have been minimum temperature issues in some school bus operations. However, some filter devices have since been proven and verified to meet these low-temperature applications. In addition, new technology to thermally regenerate filters through plug-in technology will be temperature-independent and should allow filters to be used on all model years of school buses. Pre-installation data logging is imperative to determine the proper fit between technology and operating environment.

Infrastructure. Currently, most of the school districts applying for Clean School Bus grant funds for alternative fueled buses do so because they already have alternative fueled buses and have ready access to the necessary re-fueling infrastructure. A school district that has neither alternative fueled school buses nor ready access to the infrastructure may not choose the alternative fuel bus option because of the absence of available infrastructure grants under the Clean School Bus USA Program. Some state programs, however, include infrastructure grants, such as California's Lower-Emitting School Bus Program.[19]

3. Diesel Reduction Strategies

School bus fleets are employing a variety of strategies to reduce their diesel pollution. Some districts are retrofitting their buses with DOCs, which provide a 20-40% or more reduction in particulate matter pollution. DPFs offer up to a 95% reduction in PM. If ULSD can be obtained, districts have switched over with few problems, with the exception of a few engine types whose fuel pumps have malfunctioned. Some districts have implemented CNG projects, often in conjunction with a large city or county CNG facility or the vehicle owner's own facility. Biodiesel and other fuels have been used routinely in districts with few problems.

Newer technologies, such as open or closed crankcase ventilation systems, wire mesh filters and thermally-regenerated filters (which can be used on nearly all model year vehicles) look promising. Finally, many districts are implementing idling reduction policies, which save fuel and provide health and environmental benefits. Each district chooses the diesel reduction option which best suits its own conditions, considering funding, routes, number of vehicles and other variables.

4. Incentives for the School Bus Sector

More Funding. Clearly, making funds available in the form of SEPs, grants, or other funding mechanisms seems to be the best incentive for the implementation of clean school bus projects. The need and desire for funding outstrips the availability by at least 10:1 for grant and SEP opportunities. Once the funding is available school districts become interested in implementing clean school bus projects.

Tax Incentives. A federal tax credit for the purchase of clean school buses and retrofit equipment could encourage private fleet owners to update their fleets voluntarily. Similarly, states can encourage cleaner school buses among the private sector by providing sales or property tax exemptions, and waivers of registration fees. With an estimated 140,000 school buses under private ownership, these incentives could make a significant difference in air quality.

---

[19] This is a grant program that pays for the incremental costs of purchasing new alternative fuel school buses or retrofitting certain diesel buses with exhaust aftertreatment devices. It also provides grant funding to help defray the cost of building alternative fuel infrastructure. See http://www.arb.ca.gov/msprog/schoolbus/schoolbus.htm for more information.

Many fleets, like most transit agencies, are operated by non-tax paying entities (e.g., municipalities). Income tax-related mechanisms are not effective in motivating these fleet operators. The Energy Bill (Pub L 109-49) is structured to address this issue, however. The energy bill provides an income tax credit of up to $32,000 for the purchase of alternative fuel vehicles. The Energy bill also includes tax credits for the purchase and installation of alternative fuel fueling equipment. For non-tax paying entities, the seller of the vehicles may take the tax credit, with some or all of the savings passed along to the buyer. This is an excise tax credit that can be claimed independent of the amount of excise tax paid. The SAFETEA-LU (Pub L 109-59) includes a tax credit of $0.50/gallon in the case of liquid alternative fuels and $0.50/GGE in the case of gaseous fuels for the sale of alternative fuels used in motor vehicles.

5. Other Recommendations for the Sector

In addition to incentives and funding, a number of other actions are recommended to reduce emissions from diesel school buses across the nation.

EPA's Clean School Bus USA Program should:

- Develop an education outreach program in conjunction with the national school bus transportation associations and other stakeholders to inform and educate potential grant recipients on the fundamental aspects of the program, the grant application process and the need for cleaner school bus fleets.

- Provide vehicle emission performance goals for states to consider when creating their state school bus specifications.

- Strive for geographic diversity, reaching out to smaller and less affluent school districts across the country.

- Re-evaluate any legal impediments to maintaining the EPA's emphasis on directing SEP funds toward school bus retrofit and replacement programs, since the current Clean School Bus USA Program is still a demonstration program (it is short-lived, geographically incomplete and technologically incomplete).

- Give priority to replacing the oldest buses first, especially those built before April 1, 1977 (these buses do not have to meet current safety or any emission standards), with a secondary emphasis on buses built after April 1, 1977, but before model year 1991.

- Focus on clean-up effectiveness, and the cost-effectiveness of retrofit and replacement strategies, including the effects on children's health.

- Promote strategies that achieve the lowest per-vehicle tailpipe emissions and on-bus exposures.

- Work to make sure that private contractors who own and operate school buses have equal access to program benefits, such as grants, instructional materials, technical assistance, etc.

The Clean School Bus sub-group strongly supports more funding for the Clean School Bus USA program.

B. Freight Sector Report

Ground freight transportation, the movement of goods using trucking fleets and rail, forms a solid foundation for maintaining our country's economic prosperity and competitive advantage. Moving freight accounts for 20% of all energy consumed in the transportation sector. Trucks carry about 66% of all freight shipped in the US, while rail carries about 16% (water, pipeline, and air transport account for the rest). Together, truck and rail transport consume over 35 billion gallons of diesel fuel each year. This fuel consumption produces over 350 million metric tons of carbon dioxide each year. In addition, ground freight contributes 40% of transportation-related emissions of NO_X and 30% of PM emissions.

The trucking industry transports the largest volume share of any mode of freight transportation. Corresponding to its volume share, the trucking industry is also a major contributor of air emissions from the freight sector. As shown in Figure IV.1, trucking accounted for nearly two-thirds of the freight tonnage transported in the U.S. in 2002. This volume exceeded the next largest mode of freight transportation by a factor of 3.

Figure IV.1: Modal Share of Freight Tonnage, 2002

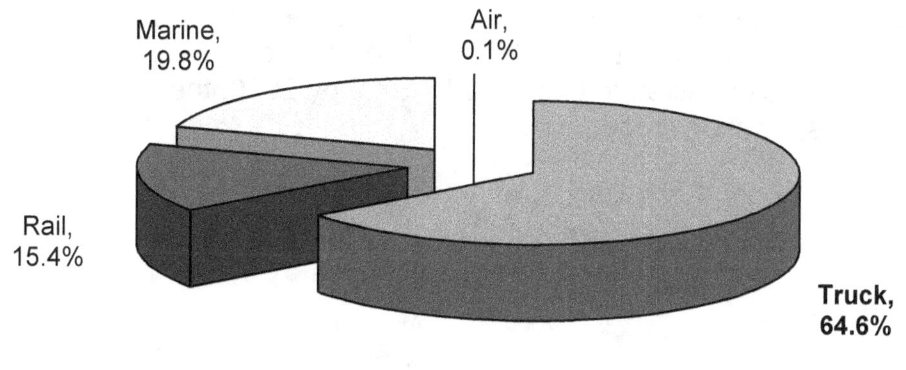

Source: Bureau of Transportation Statistics, *National Transportation Statistics 2004*.

Similarly, trucking accounted for two-thirds of the NO_X and PM emissions from freight transportation in the U.S. in 2002. As shown in Figure IV.2, NO_X and PM emissions essentially mirror the volume of freight transported from each of the respective freight transportation mode.

Figure IV.2: U.S. Freight Transportation NOx & PM-10 Emissions by Mode, 2002

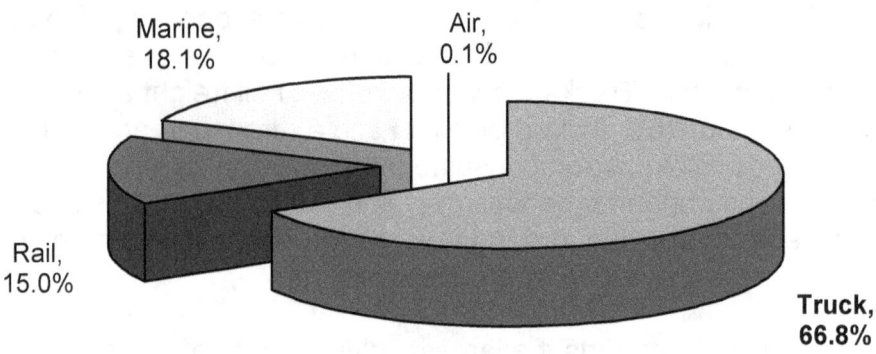

Marine, 18.1%

Air, 0.1%

Rail, 15.0%

Truck, 66.8%

Source: U.S. EPA, *National Emission Inventory.*

To account for the impact current and future engine emission and fuel standards will have on freight transportation, estimated future emissions from truck, rail, marine vessels, and air have been made. These estimates anticipate total freight emissions declining 63% by 2020.

As shown in Figure IV.3, the truck portion of total freight-related NO_X and PM_{10} emissions is expected to be cut in half over the next 15 years even though the truck's share of the freight market is expected to grow. NO_X and PM emissions from trucks are expected to decrease by 82% by 2020, the largest decrease of any freight transportation mode.

Figure IV.3: U.S. Freight Transportation NO_X & PM_{10} Emissions by Mode, 2020

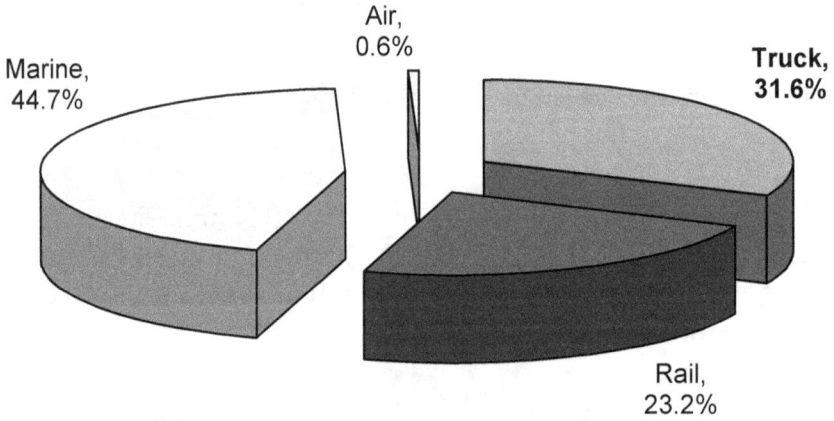

Marine, 44.7%

Air, 0.6%

Truck, 31.6%

Rail, 23.2%

Source: U.S. FHWA, *Assessing the Effects of Freight Movement on Air Quality at the National and Regional Levels,* April 2005

Demand for transport by truck and rail has dramatically increased over the past two decades, to the extent that travel currently exceeds infrastructure capacity. The EPA has established strict regulations for the trucking industry, which are expected to decrease air emissions. However, additional reductions can be realized by providing the industry with incentive-based programs, geared toward encouraging trucking companies to voluntarily increase their fuel efficiency and decrease their impact on the environment by applying emission reduction technology planned for 2007 engines to earlier model year trucks.

In this section, incentive programs are grouped into two categories: 1) programs applicable to trucking companies, drivers, and owner-operators; and 2) programs applicable to state and local government officials. Programs applicable to trucking companies, drivers, and owner-operators are those that include monetary assistance and public recognition as incentives for reducing emissions. Programs applicable to state and local governmental officials are those that include State Implementation Plan (SIP) and conformity credits as incentives for establishing assistance programs for the trucking industry.

1. Programs Applicable to Trucking in the Freight Sector

National Grants. Trucking companies and owner-operators often lack the capital to invest in emission reduction technologies or to purchasing new model year, lower emission trucks. Grant programs typically cover part or all of the initial cost of these technologies, and have proven to be effective at providing companies with incentives to use these technologies.

The Diesel Emission Reductions Act (DERA), also known as the Voinovich Bill, is by far the best national effort to achieve the legacy engine emissions improvement. These provisions have been included in the recent Energy Bill (Pub L 109-49), and will provide states with $200 million in grants for retrofitting existing diesel fleets.[20] However, for a grant program to be effective, it is essential that significant governmental funding, above and beyond that authorized through DERA, is available.

The SmartWay Transport Partnership. The SmartWay Transport Partnership is a voluntary EPA program that provides trucking companies (including owner-operators) with market-based incentives to reduce emissions. Shippers commit to decrease their environmental footprint and to use SmartWay carriers. Carriers (those who move goods for shippers) commit to adopt technologies and strategies that improve fuel efficiency, save money, and reduce their emissions. As a result, carriers are encouraged to continue to improve their environmental performance so that their company is more attractive to potential shippers that may hire them. This provides trucking companies with a direct incentive to voluntarily reduce their

---

[20] "Bush signs Energy Bill, Clean Diesel Provisions into Law." *Diesel Technology Forum.* Website: http://www.dieselforum.org/ accessed September 29, 2005.

emissions. All Partners receive recognition for their efforts through press releases, publications, the SmartWay website, etc. The Partnership therefore represents a win-win-win for participants, the public, and the environment.

Innovative Technology Bundles That Put Money in a Truck Owner's Pocket. One of the toughest challenges to overcome is the fact that most retrofit technologies (PM filters, oxidation catalysts, etc.) generally provide little or no intrinsic economic benefit to the user. Therefore, these emission reduction programs/incentives designed for diesel powered fleets (especially those aimed at private companies) are facing an uphill battle from the beginning. However, if a program or incentive were developed that provided direct economic benefit, then companies and organizations would develop interest at a much greater rate. In the freight sector, trucks are the largest consumer of diesel fuel. With a single long-haul truck capable of consuming over 17,000 gallons per year, a fuel economy improvement of just 10% could provide over $4,000 in savings each year (assuming a fuel price of $2.50 per gallon). Such a creative program could use these savings to pay for additional emission control. The program would need to bundle innovative fuel saving technologies along with traditional retrofit technologies. The program will require innovative capitalization methods, innovative loan structures, or innovative tax waiver processes to help companies overcome the initial capital investment of the technology bundle.

- Technology Bundling. For technology bundling to be effective, it is essential that the bundle contain a combination of highly efficient, fuel saving technology and an emission control technology. Types of innovative fuel saving technologies that should be included in this bundle, and their associated fuel savings, are:
  - Idling Control Technologies: 6-10% fuel savings[21]
  - Super single tires with aluminum wheels: 4-10% fuel savings[22]
  - Improved Aerodynamics: 5-7% fuel savings

Along with the fuel saving technologies, a company should choose to use an oxidation catalyst, PM filter, or other PM emission control device. The key is that the technology should be carefully selected so that the upgraded truck will provide the owner with a net economic benefit.

- Innovative Capitalization and Loan Programs. Most small to medium sized trucking companies do not have the capital to invest in these technology bundles. Therefore, innovative financial programs are needed to assist companies upgrade their trucks. Currently, two states (Arkansas and Minnesota) have innovative loan programs that provide capital to trucking fleets for "SmartWay Upgrade Kits" that combine fuel saving technology with emission reduction technology. These programs are unique because they not only

---

[21] EPA, Draft Report for Review, Industry Options for Improving Ground Freight Fuel Efficiency, 2002; Lim, H. Study of Exhaust Emissions from Idling Heavy Duty Diesel Trucks and Commercially Available Idle-Reduction Devices, 2003, SAE Paper No. 2003-01-0288.
[22] Estimates based on OEM data, fleet data, and EPA preliminary testing. EPA is currently conducting additional fuel economy and emissions testing on these products.

provide companies with an incentive to purchase retrofit technologies, but they also allow the companies to immediately become more profitable.

The following example demonstrates the profitability of this type of loan program[23]: Consider a $14,300 technology bundle of an: auxiliary power unit, wide base tires and wheels, trailer aerodynamics, and an oxidation catalyst

Monthly loan payment:	$ 400
Monthly fuel savings:	$ 600
Monthly profit:	$ 200 Money in an owner's pocket

After three years, profits for the company jump to $600 per month.

Even with a particulate filter, at a total cost of $19,400, this technology bundle is still profitable.

Monthly loan payment:	$ 580
Monthly fuel savings:	$ 600
Monthly profit:	$ 20 Money in an owner's pocket

After three years, profits for the company jump to $600 per month.

Extended Privilege Packages. More than ever, the trucking industry is under pressure to deliver faster, to deliver within very tight delivery schedules, and to work within just-in-time delivery constraints. These factors, coupled with the expected growth in freight movement over the next decade provide some opportunity to minimize the "hassle" associated with moving goods across the country. Those companies and organizations that agree to participate in emission reduction programs would be granted certain privileges that would improve the company's throughput and improve their ability to deliver on time. Extended privilege packages could include, but are not limited to:
- Use of high occupancy vehicle (HOV) lanes;
- Priority parking;
- Easy access to loading docks (avoiding wait times);
- Weigh station and inspection flexibility;
- Tolling leniency; and
- Efficient border crossing systems.

Some of these privileges must be developed and implemented by state or local governments, while others may require federal government oversight. In some cases, extended privileges can be developed and implemented by private shipping companies (e.g., maintaining a loading dock bay reserved only for low emission trucks).

---

[23] Assuming annual fuel consumption of 18,000 gallons and fuel cost of $2.50/gal, 2,400 hours idling, and a 36 month loan at 4.8% APR

Tax Incentives and Waivers. Companies and owner-operators are currently charged an excise tax for several innovative technologies on the market today. This tax hinders them from purchasing these technologies. An excise tax waiver would remove this barrier and provide an extra incentive for companies to purchase efficient technologies. An income tax waiver, federal and state, for the incremental capital purchase, will greatly improve the appeal of such a program.

Another barrier in the marketplace is the application of weight limitations for add-on technologies, such as auxiliary power units (APUs) and some retrofit devices. A weight waiver should be applied to these products so that trucking companies can continue to carry maximum loads if they decide to invest in emission control technology. For example, the Energy Bill (Pub L 109-49) includes a 400-pound weight exemption for APUs.

Truck Labeling. Many trucking companies (especially those companies or fleets that are recognized by the public) are interested in marketing their environmental progress and believe that one of the most cost effective ways they could do this is with their trucks. A truck labeling program would allow trucking companies, owner-operators, and any company with a trucking fleet to showcase those trucks that have innovative, emission reduction technologies. Only those trucks that are equipped with sophisticated, proven emission control technology would be able to display the label or logo. Specific emissions thresholds must be established to create a "level playing field" for all companies so that when one sees a truck with a label or logo, it is clear that it is a low- emission truck. The SmartWay Transport Partnership is currently developing a truck labeling program.

2. Programs Applicable to State and Local Government Officials

SIP and Conformity Credits. Although each of the strategies discussed above create incentives for emissions reductions from freight, state and local air quality agencies have had difficulty claiming SIP and/or Conformity credits for these reductions because most long-haul trucks do not operate primarily in a single area. Instead these trucks operate inter-state, regionally, or nationally. Creating a program or air quality guidance that describes how emission reductions from long-haul trucks could be credited in SIPs would serve as a significant incentive for states and local governments to, in turn, create programs offering incentives as described above.

Fuel Efficiency/Emissions Reductions. The SmartWay Transport Partnership's technology verification program is studying the relationship between fuel efficiency and emissions reductions. Trucking companies and owner-operators are interested in increasing their fuel efficiency because it will reduce their fuel consumption, save money, and reduce emissions. State agencies and local officials are interested in emissions reductions for human health protection and SIP compliance. Therefore, it is important to be able to quantify the emissions reductions that result from increased fuel efficiency. The relationship between increased fuel efficiency and decreased emissions serves as an incentive for state and local governments to form assistance programs for the trucking industry.

3. Recommendations

The following action items are recommended for EPA to consider in developing a diesel emissions reduction strategy for the trucking sector:

- EPA should create a national capitalization program designed to provide capital at attractive market rates and terms for trucking companies and fleets of all sizes. Additionally, EPA should work with private lending institutions to create innovative capitalization programs that include technology bundling. EPA should explore the use of income tax waivers for such qualifying capital purchases.

- In addition to federal leadership, aggressive coordinated leadership is needed from all parties including states, NGOs and trade associations to achieve Congressional and state-legislative support to implement high dollar programs for government grants, tax incentives/waivers and/or rebates structured both for non-profit organizations and for-profit companies.

- EPA should explore implementing the loan programs, tax incentives, and labeling programs for hybrids. Some members also thought extended privilege packages would be useful.

- Use EPA's SmartWay Transport Partnership to continue to increase the demand for cleaner, more efficient freight delivery services.

- EPA should test and verify the effectiveness of innovative technology bundles that include fuel saving and emission reduction technologies to determine the emissions reduction potential and return on investment scenarios. EPA should then publicize and market the results to states, local governments, and trucking companies.

- EPA should work with states and local agencies to expand the number of innovative loan programs that provide capital to trucking fleets for "SmartWay Upgrade Kits." Currently only Arkansas and Minnesota have such programs.

- EPA should work with states and local authorities, as well as private companies to explore the development of extended privilege packages for trucking companies.

- EPA should continue its efforts to create weight waivers for innovative technology that can be added to trucks.

- EPA should develop criteria identifying the emission control thresholds for a SmartWay truck and should create a program that allows trucking companies to label qualifying trucks in their fleet that meet the emission control thresholds.

- EPA should continue to study the relationship between fuel efficiency and emissions reductions and should identify as many technologies as possible that both reduce emissions and save fuel. For those technologies that both save fuel and reduce emissions, EPA should prepare formal air quality guidance that will allow states to credit emission reductions from fuel efficiency technologies in SIPs and conformity. The guidance should identify methods by which several nonattainment areas could receive credit as a result of retrofitted long haul trucks passing through the area.

- EPA should determine how to apportion air quality benefits across multiple jurisdictions based on fuel consumption and fuel tax reporting requirements and other measures (e.g., satellite tracking). In addition, EPA should explore technology-driven apportionment programs that would potentially facilitate the involvement of national fleets in a national retrofit program while still allowing for the calculation of local air quality benefits.

- EPA should work closely with DOE to undertake research and development of new technologies to conserve fuel and reduce emissions. DOE needs to continue to develop and test technologies.

One issue on which consensus was not reached was whether EPA should evaluate the feasibility of mobile-to-stationary source trading credits for shippers.

C. The Marine Ports Sector Report

The United States has 185 deep-draft seaports located along the mainland coasts of the Atlantic, Pacific, Gulf of Mexico and Great Lakes, as well as in Alaska, Hawaii, Puerto Rico, Guam and the U.S. Virgin Islands. Together these ports provide approximately 3,200 cargo and passenger handling facilities, according to the U.S. Coast Guard. Most of these deep-draft ports are controlled by public agencies that are arms of state or local governments or special districts, commonly referred to as public port authorities. Additional in-land ports are on our nation's rivers and waterways.

Commercial seaports handle a variety of cargoes, including bulk (loose) cargo, breakbulk commodities (packages such as bundles, crates, barrels and pallets), liquid bulk (such as petroleum), roll-on/roll-off cargo (also called "RO/RO," which includes farm equipment, automobiles, and military deployment equipment), and containerized cargo (steel boxes measured in 20-foot equivalent units or TEUs). Cargo generally enters a port through a marine terminal, and several terminals typically constitute a port.

Cargo volumes through deep-draft seaports are growing rapidly. The total volume of foreign trade moving through U.S. ports is expected to double 1996 levels by the year 2020.[24] It should also be noted that many commercial seaports serve the cruise passenger industry, which is also growing rapidly. From 2002 to 2003, the number of U.S. passengers cruising increased 9.4%.

Over 30 of the largest ports are located in areas that are designated as nonattainment for the NAAQS for either PM or ozone or both, and many of these are in areas that are projected to continue to be in nonattainment after many of EPA's rulemakings take effect. Others are located in NAAQS maintenance areas or where air quality levels are close to the health standards. Emission reductions from port operations in these areas will contribute to continued compliance with the NAAQS. Many ports and their surrounding communities have concerns with air toxics, and diesel particulate matter has emerged as an important public health threat. As cargo volumes continue to grow, more vessels, cargo-handling equipment, trucks, and trains will be needed to accommodate this increased trade. Mobile source emissions associated with goods movement are having an increasing effect on adjacent communities. Many opportunities exist to reduce emissions from diesel engines in and around port communities.

---

[24] In 2002, ports invested nearly $1.7 billion to update and modernize their facilities, almost equaling the record set in 2001, including: $140 million for general cargo; about $942 million in investments related to containers; $241 million on infrastructure improvements. During the 5-year period between 2003 and 2007, public ports predict they will spend $10.4 billion (a record level), compared to actual expenditures of $7 billion between 1998 and 2002. (Source: Source – Maritime Administration, U.S. Department of Transportation, "*United States Port Development Expenditure Report,*" May 2004.

Seaports nationwide invest substantial resources in infrastructure, technology, and operational procedures that increase efficiency and decrease emissions per passenger or unit of freight transported. Major development projects include substantial investments in environmental projects, including air quality projects that would not otherwise be fiscally possible. Major seaports are actively engaged in developing and implementing air pollution prevention projects.

Diesel engines are in frequent use in almost all port activities. They power the ocean-going vessels that carry cargo as well as passengers on cruise lines from port to port, and smaller harbor craft such as tugboats and ferries. They power the cargo-handling equipment used to load and unload containers from ship to shore (cranes) and within the terminal itself (such as rubber-tired gantry cranes and yard hostlers). Diesel engines also power the trains and trucks that move containers into and out of the marine terminals.

Many different entities own and operate the diesel equipment that is present at ports. Port authority operations can be categorized as follows:
- Operating ports directly own and operate cargo-handling equipment (the Port of Boston is an example of an operating port);
- Landlord ports, the most prevalent in the U.S., lease property and/or equipment to terminal operation companies that own and operate the dockside equipment and are responsible for all operations such as loading and unloading of vessels (major port authorities such as Los Angeles, Long Beach, Seattle, and New York/New Jersey are examples); and
- Hybrid ports are an amalgam of the operating and landlord/tenant ports in that they both operate their own on-dock equipment as well as lease land to terminal operators (the Port of Baltimore is an example).

With the rapid growth of containerized cargo and passenger traffic over the past few decades, most major ports now have a significant portion of their properties dedicated to container terminals and cruise lines. Containerized freight operations by far use diesel powered equipment more intensively than other types of freight. Because of the nature of container terminal operations and the growth in volume of waterborne cargoes, ship calls are more frequent (and larger), truck visits are more frequent, and cargo-handling equipment usage is increased; thereby, typically generating more diesel fuel emissions than at other kinds of terminals. While the following discussion focuses on container operations as examples and quantitative data from the Port of Los Angeles which recently completed a comprehensive emission inventory, it is important that incentives and voluntary reduction programs be designed for all ports across the country.

The terminal operator industry has undergone significant consolidation over the past few decades. Today probably a little more than a couple of dozen terminal operating companies are still operating in the U.S. Furthermore, many of today's terminal operators are subsidiaries of shipping companies and provide this service to their affiliated companies as well as to other shipping companies. These terminal operations companies typically operate at arms length from their affiliated shipping and trucking companies. There remain a handful of independent terminal operators that still hold a significant share of the market. Whether owned by the port authority or a terminal operator company, based on the Port of Los Angeles study, cargo handling equipment constitutes approximately 10% of the port NO_X and 13% of the port direct $PM_{2.5}$ emissions. For example, roughly 1,000 pieces of cargo handling equipment were at the Port of Los Angeles. The main types of cargo handling equipment at ports include yard tractors, cranes, forklifts, and top and side handlers.

Another major source of diesel emissions near ports emanates from the trucks that call on ports, which are typically older models. These short-haul or "drayage" trucks are usually independently owned and operated by small, economically struggling companies. The owner of the cargo may contract for delivery services through a trucking service company. Trucks can form bottlenecks at port terminal entrance gates, where they may idle. A single port complex can receive thousands of trucks entering and leaving on a typical day. For example, more than 32,200 diesel truck trips occur in and out of the Port of Los Angeles and Port of Long Beach complex, which is North America's busiest port complex, and a large percentage are pre-1984 model years that were not subject to today's emission control requirements. For example, in the Port of Los Angeles, heavy-duty trucks currently calling on major container ports emit about 23% of the port NO_X and about 9% of the port directly emitted $PM_{2.5}$. These figures are subject to uncertainties depending on where one considers the boundary for port-related truck traffic.

More than three-quarters of all train traffic transports containers, and most of these trains are traveling to or from marine ports. The rail category includes both line haul (see the freight sector of this report) and switching. On-dock rail is used by some ports to efficiently move cargo directly from ships to rail lines. On-dock rail cannot be efficiently utilized at some ports due to space limitations on terminals, ownership of lines, and other factors. Rail contributes approximately 13% of the port NO_X emissions and 6% of the port directly-emitted $PM_{2.5}$ at the Port of Los Angeles.

Marine vessels, including harbor craft (e.g., tugboats, towboats, and ferries) and large ocean-going vessels (e.g., container ships, tankers, and cruise ships), emit about 54% of the port NO_X and 72% of the directly-emitted $PM_{2.5}$ at the Port of Los Angeles. Ocean-going vessels alone accounted for 53% of the port directly-emitted $PM_{2.5}$. Container ship traffic to and from the US doubled between 1990 and 2001 and the rate of increase is expected to continue. Figures IV.4 and IV.5 represent these figures.

While many port authorities and terminal operators have been proactive in implementing programs to reduce emissions from terminal operations, many significant opportunities still exist within a typical marine terminal. Ports have invested in air pollution prevention projects at the same time they were coping with substantial post-9/11 economic stress. These improvements must be achieved while ports face a number of key challenges. For example, port authorities are subject to mandates for Homeland Security measures at seaports. Ports are also concerned about operational reliability, the need to manage risks that might impede their ability to transfer cargo in a timely manner. Ports are also highly competitive with each other in a dynamic market where freight owners and terminal operators will select the port with the greatest efficiencies and lowest cost that best meets their business requirements. Furthermore, the different regions where seaports are located have very diverse air quality challenges. Each port needs to work closely with their local and state air agencies in setting pollutant priorities to assure their voluntary air quality investments are aligned with local needs.

Figure IV.4

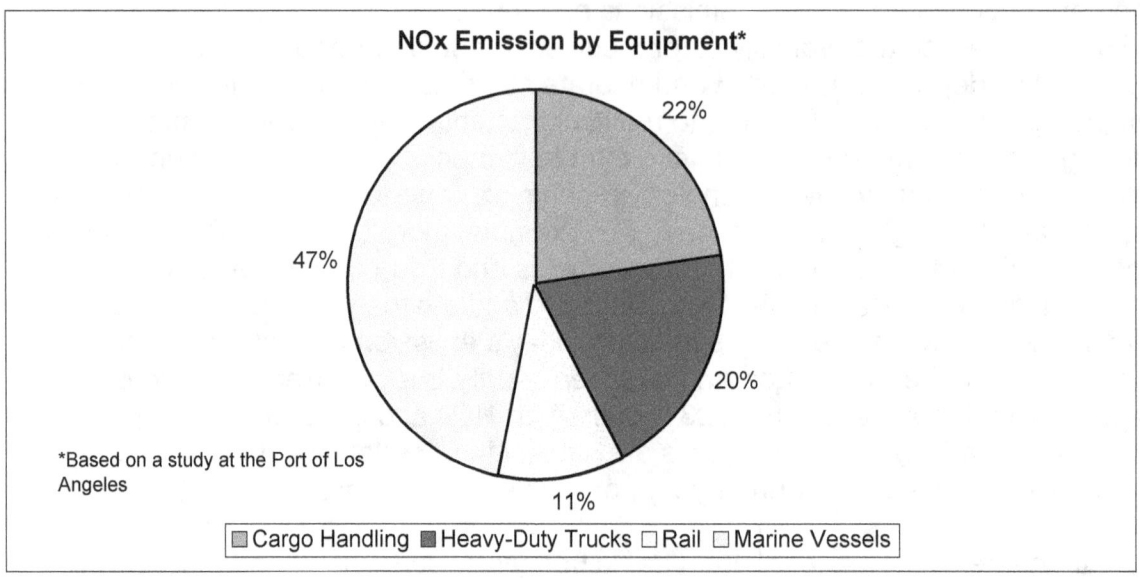

Figure IV.5

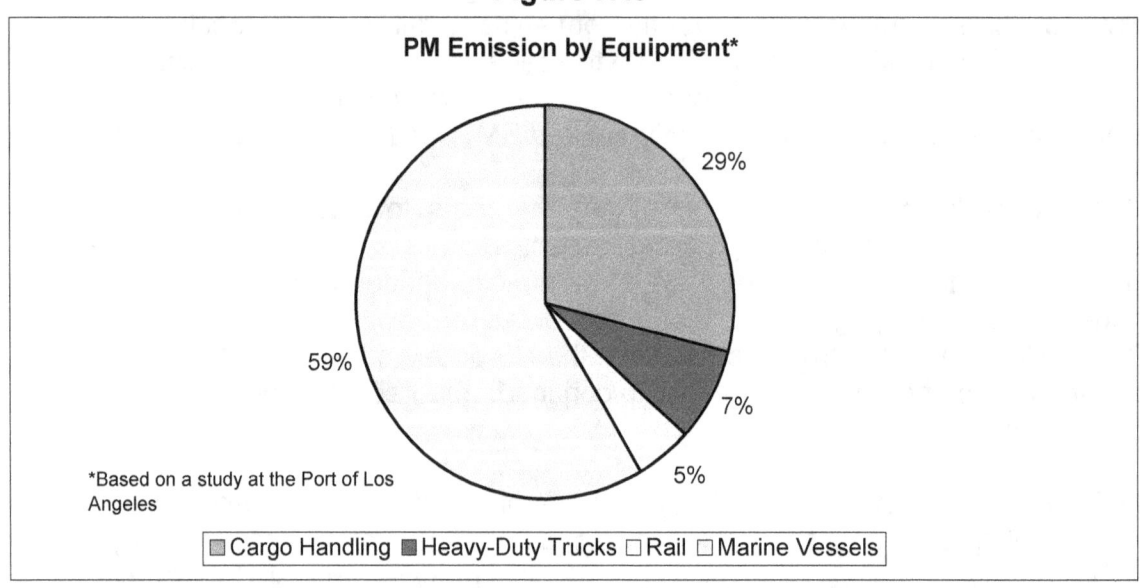

While ports are successfully demonstrating a wide array of diesel emission reduction strategies, a need exists to continue to develop new techniques and to share best practices among ports. Given the diversity of operations at ports, different entities, each with a unique business model, will likely take different approaches to reduce air pollution, suggesting that a diversity of incentives and technologies may be needed to achieve voluntary reductions.

Appendix B lists possible diesel emission reduction strategies which are categorized by switching to cleaner fuels, installing retrofit devices, implementing operational strategies, and repowering engines or replacing engines or equipment. Many ports are taking a leadership role in switching to cleaner fuels, such as using 500 ppm highway grade diesel in nonroad equipment or ULSD in advance of the required deadlines. Several major terminal operators have favored replacement options because of their need for reliability and having engine manufacturers cover all warranty claims. Trucking companies may favor options that save fuel (e.g., gate improvements and anti-idling). Switching to cleaner fuels (e.g., ULSD in nonroad equipment) is a very promising strategy for reducing emissions.

1. Challenges

Perceived or real barriers may exist that must be overcome with carefully crafted incentives that accommodate the differing business models at ports. With pragmatic incentive packages, entities operating at ports would be more likely to voluntarily adopt effective emission reduction strategies. Towards this end, there are economic, technological, educational, and programmatic challenges for ports in implementing emissions reductions, as detailed below:

Economic. Ports are a collection of competitive enterprises where bottom line concerns are paramount. Cost of technologies and cleaner fuels, reliability (as down time can be costly both to port authorities and terminal operators and to ships and trucks who call on the ports), and access to capital (for equipment modernization) may be issues. In addition, ports are facing Homeland Security mandates that often require resources, but also can provide additional opportunities for emission reductions. Grant application deadlines may be out of sync with port business cycles or the administrative burdens may be high for the relatively small fraction of a project that a grant may provide. Grant funding is also limited and, therefore, may not be able to fund all merit-worthy projects. Some small businesses, such as independent truckers, may be uncomfortable with federal or state grant process and may work best with rebates offered through truck dealers or retrofit and electrification programs administered by ports or other local agencies that could simplify the process, such as currently done in Los Angeles and Long Beach through Gateway City funding. As waterborne freight increases, port operations across the country face pressure to move more cargo with limited resources. In some cases, addressing air quality issues can aid ports in meeting efficiency demands, and these options should be pursued. Ports located in states and municipalities that are working to reduce greenhouse gas emissions have added reasons to favor strategies that increase efficiency and reduce fuel consumption. Competition among ports and enterprises is also an issue. Moving forward, voluntary incentives that assist ports in becoming more efficient and productive in a competitive market while reducing emissions will be desirable. Failure to do so could merely transfer the air emissions and the associated economic benefit to another community without solving the problem. Ports, especially those serving common markets, could implement some provisions collaboratively to minimize these problems.

Technological. Because ports and terminal operators feel they cannot risk an interruption in their business operation, they are hesitant to adopt new technologies that are not verified or certified or do not have a reliable track record. While new technologies are being developed and tested, manufacturers offer only a limited number of verified/certified technological options with established track records for ports, especially for nonroad applications and NO_x controls. Technology demonstrations and more widely available cleaner fuels are needed. The incremental cost of cleaner technologies when not offset by fuel savings or maintenance improvements or other business case reasons to adopt the strategy is a barrier. Also, technologies (engines or retrofit devices) in high demand may not be available without substantial lead time.

Educational. Challenges also include keeping busy port administrators, terminal operators and fleet managers current on air quality issues, public health concerns related to air quality and the complex range of emissions reducing options. Educational needs include sharing best practices and lessons learned among port enterprises. Ports face complex jurisdictional issues, with a myriad of federal, state, and local agencies. Coordinating with these agencies, with companies who do business at ports, and NGOs takes time and an educational process of all parties. Therefore providing ports with the tools and technologies to employ effective emissions reducing projects and to build collaborative relationships is also needed.

Programmatic. Ports across the country are diverse—each with different needs, management structure, air quality issues and business operations. To accommodate the diversity in ports and enterprises at ports, flexibility and a suite of incentives will be needed.

2. Incentives

Since no single incentive will be able to eliminate all barriers to reducing diesel emissions, a suite of solutions is the best strategy to address each of the barriers above.
A number of incentives exist to encourage public port authorities and other companies that own or operate equipment in and around U.S. seaports, to voluntarily reduce air emissions through one of the technological or operational methods identified. However, the operating structures of public port authorities vary widely, and a number of different companies or organizations may own or operate diesel equipment at a given commercial port.

Different incentives offer different levels of appeal to different fleet owners. Because of the frequently-cited cost barrier, many incentives identified are monetary. In evaluating incentives, this work group has sought to identify solutions that are feasible, functional, and flexible:

44

- FEASIBLE – Well-crafted incentives are needed to overcome barriers and likely to spur voluntary action by public port authorities and other entities that own and operate fleets in and around U.S. commercial seaports.
- FUNCTIONAL – Incentives will encourage implementation of emission reduction strategies that yield meaningful air quality improvements at local and regional levels.
- FLEXIBLE – Incentives accommodate the different types of operating structures, cargoes, equipment in use, and air quality challenges of the diverse U.S. public port industry and are available to all ports regardless of attainment status.

Grants. Grants have been identified as an important incentive to overcome the cost barrier for strategies that don't offer strong business case support. Because of the scale of many ports and the high cost of the diesel equipment in use, grant amounts need to be large enough to overcome perceived administrative barriers of applying for and overseeing grants. For example, the No Net Increase report from the Port of Los Angeles preliminary estimate for holding the line on diesel emissions is between $11.6 and $15.7 billion for a single major port, and this would result in $28 billion in public health benefit.

Tax Incentives. Tax incentives are appealing to many private companies (such as terminal operators, and tug and tow companies) because they have no application deadline, and allow firms to apply on their own schedules without fear that incentive funds will be exhausted. However, public port authorities that pay no taxes cannot take advantage of tax incentives. With tax incentives, unless they are very narrowly targeted, it may be more difficult for the government to direct resources at the diesel emissions of greatest concern or to make changes to the program. To be effective, tax incentives, whether in the form of a tax deduction or a tax credit, must be set high enough to induce firms to make improvements to their diesel equipment that they otherwise would not do. Since efficiency gains are generally not realized from the retrofit of diesel equipment or use of alternative fuels, and therefore no return on investment, companies might not be motivated by a tax incentive of less than 100%. However, a tax incentive of less than 100% could be successful if applied for a fleet modernization strategy.

Loan Programs and Rebates. Especially appealing to small businesses, loan programs provide flexible capital to fund emission reductions efforts. These incentives may be appropriate for trucking firms serving ports.

Contract or Lease Requirements. Contract or lease requirement effectively mandate emission standards. If employed on a port-by-port basis, they may put ports at a competitive disadvantage with one another, with private terminal operators, or others affected by the contract or lease. Also, contract or lease requirements may negatively impact small businesses, as small companies may not have the ability to finance the equipment upgrades necessary to win work under a contract or lease specification. Additionally, the long leases at many terminals and, thus, infrequent opportunities to negotiate new lease terms hinder the effectiveness of lease specifications in achieving port-wide emission reductions. However, port expansions may provide opportunities for this incentive to be used, as has been the case in the Ports of Long Beach and Los Angeles.

Recognition/Awards: Companies are increasingly finding that it makes good business sense to proactively embrace environmental stewardship rather than react to government regulation or a negative public image. Government can help encourage these steps by offering guidance, education, and recognition. However, while recognition and awards programs provide positive incentives for action, they do not address some of the key barriers to action, such as implementation cost.

Regulatory Credits. Many public port authorities have identified barriers to voluntary action within the regulatory process. Offering ports the ability to claim site-specific emissions credits, either within a SIP, a NEPA process, or during a general conformity rulemaking, is an incentive. Governmental entities and public port authorities can be motivated by SIP and conformity credits to reduce diesel emissions. Without a way to bank site-specific credits, ports might not make early reductions that they feel would be needed for later expansions or projects. Any credit program should ensure the credits are surplus, verifiable, quantifiable and enforceable. In addition, record keeping and monitoring for credits must be reasonable to avoid creating another barrier to early reductions. In this regard public port authorities and others have requested guidance and recognition for claiming credits and an ability to bank them for future use.

3. Recommendations

Solutions differ from one port to another. EPA should assemble a suite of solutions recognizing that different enterprises will have different drivers for emission reductions. These solutions will be implemented on a local, port-specific basis.

- Grants. EPA should work through its budget process to recommend grant programs be offered to demonstrate technologies and to encourage the routine adoption of cost-effective diesel emission reduction techniques. Both port authorities and private companies who do business at ports are interested in receiving grants. Because of the constraints on the Federal level to award grants to private entities, EPA should also work with stakeholders to create a model state program and educate states about how they can use their fee authority to create a program like California's Carl Moyer Program or Texas's TERP to provide grants to retrofit or modernize port-related equipment.

- Tax incentives. EPA should work with the IRS to develop a model tax credit for companies (marine terminal operators, vendors who lease diesel equipment, railways and/or trucking firms) who endeavor to modernize their fleets to achieve early emission reductions. Favorable depreciation provisions for tax purposes should be included for the differential cost of equipment voluntarily purchased to reduce air emissions.

- Low Interest Loans/Rebates. EPA should identify financial institutions that could work together in an area to provide low interest loans (or rebates through

46

authorized dealers) for independent owner/operators to upgrade engines or purchase a package of diesel emissions reduction/fuel savings technologies. This approach may be applicable for some terminal operators and leasing companies.

- Freight Infrastructure. EPA should coordinate with DOT (MARAD and FHW) and Homeland Security to start addressing major infrastructure support needs to accommodate the projected growth in waterborne freight and global trade trends in an environmentally beneficial way that improves air quality. EPA could facilitate an analysis of air quality impact of options.

- Credits. EPA should work with stakeholders to develop guidance for quantifying and claiming regulatory credits that are surplus, verifiable, quantifiable and enforceable, including a way to bank credits from early voluntary mobile source diesel emissions reductions projects at a discounted rate against future needs,

- Recognition. EPA should create a national award or recognition program for port authorities and other entities that operate at ports. EPA should promote the visibility of the National Clean Diesel Campaign and ports contribution to the effort.

- Sharing Best Practices. EPA should develop educational materials and tools to continue the education and coalition-building that has become the cornerstone of voluntary efforts to encourage diesel emission reduction activities at ports. Programs could include case studies; best practices; technical information in the form of print, web and interactive workshops; regional collaborative; and local on-going forums.

- Technology Verification. EPA should enhance its verification program and work with manufacturers and fuel suppliers to ensure adequate emission control strategies are available.

- Emissions Inventory. Encourage port authorities and other stakeholders to quantify emissions inventories voluntarily. EPA should work with stakeholders to develop emissions inventory guidance.

- Evaluation. Six months after the sunset of the Clean Diesel and Retrofit Working Group, EPA should evaluate its progress with the MSTRS.

D. Construction Sector Report

The construction industry operates in every state and employs more than seven million workers, accounting for more than 6% of the private non-farm workforce.[25] In 2004, the value of construction put in place totaled $1.03 trillion,[26] or nearly 9% of gross domestic product.[27] While they therefore play an important role in the U.S. economy, most construction contractors are small, low-margin businesses.

The industry uses more than 2 million pieces of diesel-powered nonroad equipment, which vary considerably more than highway vehicles their in size, configuration, and applications. Much of this equipment has a long operational life, often lasting more than 25-30 years. Given the magnitude of the industry, the types of vehicles employed and the proximity of construction work to population centers in many cases, construction vehicles impact air quality. According to EPA models, in 2005 construction equipment generates roughly 32% of all land-based nonroad NO_x emissions and more than 37% of land-based PM_{10}. Compared with heavy duty highway vehicles and automobiles, nonroad equipment emits more pollution and has less stringent emissions standards for comparable model years. For example, a bulldozer engine can emit as much particulate matter as more than 500 cars.

Dividing the value of all construction projects among property owners, and then listing these groups of projects in descending amounts of equipment used, yields the following:
- Public projects (roads, other public works, and public buildings) accounted for $229 billion (22%) in 2004;
- Private nonresidential projects accounted for $235 billion (23%);
- Private multi-family accounted for $38 billion (4%);
- Private single-family accounted for $378 billion (37%); and
- Private residential improvements accounted for $147 billion (14%).

Private construction companies perform most public construction using equipment that they own or lease, or rent for a short term. Private companies own roughly 93% of all new diesel-powered construction equipment, equal to 90% of the value of all such equipment.[28] Many contractors, especially small businesses, rent or lease equipment, so incentives are needed for leasing companies as well.

Although the industry's total employment and output are large, the typical construction company is very small. Of the roughly 700,000 construction firms with

---

[25] Bureau of Labor Statistics, U.S. Department of Labor, *Employment Situation*, www.bls.gov/ces/home.htm.
[26] Census Bureau, U.S. Department of Commerce, *Value of Construction Put in Place*, www.census.gov/constructionspending.
[27] Bureau of Economic Analysis, U.S. Department of Commerce, *Gross Domestic Product*, www.bea.gov.
[28] Manfredi Associates, from government and private sources.

48

employees, 92% have fewer than 20 employees. An additional two million businesses, mainly sole proprietorships, have no employees.

1. Diesel Emissions Reduction Technology Strategies

Approximately 2.1 million pieces of nonroad construction equipment are currently in use. EPA has been phasing in engine emissions standards for new model years and certain horsepower classes since 1996. The term "tier level" refers to the emission standards that a particular engine meets with tier 1 standards being the first or earliest set of emissions standards and tier 3 being the standards that new engines are meeting today. The strictest standards, tier 4, will phase in over the next decade. The higher the tier level, the cleaner the engine.

Of the more than 2 million engines that the construction industry uses, about 31% (or 650,000 pieces of equipment) have engines manufactured before any emissions standards took effect and, therefore, have no emission controls.[29] Currently, the retrofit technologies and repowering options for reducing the emissions from these older engines are limited. Early replacement is another but costly option.

Approximately 36% of construction equipment contains basic engine based emissions controls and meets EPA's tier 1 level and roughly 28% of equipment meets tier 2 levels. Only an estimated 5% of construction equipment meets EPA's current standard at the tier 3 emissions level. Appendix C contains more detailed information.

Strategies to reduce pollution from construction equipment include retrofitting with pollution controls, replacing or repowering older engines to a higher tier, using cleaner fuels, reducing idling time, and proper maintenance. Compared with highway engines, challenges to retrofitting construction equipment with pollution controls are unique. Retrofit technologies need to address issues like extended idle and/or low speed operation periods, fuel quality (including sulfur levels), vibration, high levels of fugitive dust, space limitations, and visibility are unique to this sector and require additional attention when retrofit technologies are being considered. Older engines may also have undesirable NO_X/PM ratios for use with retrofit technologies. In some cases, early engine or vehicle replacements are more cost effective, in at least the long run than the application of a retrofit technology. Proper maintenance and effective repair are the initial keys to achieving cleaner engines followed by cleaner fuels and aftertreatment devices and systems. DOCs and DPFs that are specifically designed for construction equipment will also help meet the emissions reduction goals in this sector. While not in wide use in the U.S., Switzerland has thousands of pieces of construction equipment retrofit with DPFs and will have 100 % of its construction fleet retrofit with in a few years. In the short term, idling controls, DOC installations, DPFs, crankcase controls, engine upgrades, and cleaner fuels with lower sulfur levels are among the easiest to implement and will likely be the predominant choice until SCR, NO_X absorbers, and EGR are fully optimized for application to construction equipment. The maturity of these systems

---

[29] Environmental Protection Agency, *Nonroad Model,* www.epa.gov/otaq/nonrdmdl.htm.

is expected to lag behind the systems intended for highway vehicles by several years.

2. Considerations for Designing Incentives

Effectively encouraging construction companies to voluntarily reduce emissions from existing diesel-powered construction equipment requires striking a balance among a mix of business, economic, technical, commercial, and factors including health, air quality, outreach and education. These factors are discussed below. No attempt has been made to prioritize them.

Health and Air Quality. Construction equipment varies greatly in the frequency and intensity of its use and therefore in the amount and type of pollution it emits. Public and occupational exposure to emissions from such equipment is dependent upon a variety of factors, including the location, working hours, and equipment mix used for any particular project.

Incentive programs should be designed to maximize environmental benefits. To ensure this, incentive programs should target areas of high ambient pollution, personal exposure to diesel pollutants, equipment that is most likely to contribute to high pollution levels or exposures, and the categories of equipment that are most likely to benefit from retrofit strategies or technologies.

Business and Economic. To many construction companies retrofit technologies have little intrinsic economic benefit and instead may the increase the cost/risk of doing business. Costs associated with cleaner equipment include not only the purchase price but also installation costs; the cost of owner's/managers' time in becoming familiar with alternative retrofit technologies and the terms under which they can avail themselves of incentives; the cost of overtime, substitute equipment rental, or foregone revenue from idling the equipment to install a retrofit technology; and the risk that further costs will be incurred for maintenance and training relating to a new technology. For these reasons, financial assistance needs to be great enough to cover at least the majority of the costs of the use of a retrofit technology when no economic benefit exists to the equipment owner. Even if an incentive does compensate equipment owners for most or all of these costs, policy makers need to recognize that equipment owners are likely to consider the total costs before deciding whether to adopt a retrofit technology.

The income and property tax implications of incentives also have a bearing on their effectiveness of the incentive on construction equipment business owners. For example, if not handled carefully, providing "free" retrofit technology (for example, through a grant payment covering the cost of the technology) may actually create a tax liability for the equipment owner accepting the "free" technology.

Different businesses will weigh these considerations differently. The conditions placed on an incentive will affect the likelihood that it alters the competitive situation between large and small owners, or established and new firms. For instance, large firms may be in a better position to absorb the costs of learning about and applying for incentives, while loan programs might confer an advantage to firms that already have a credit history. Other programs might include size or location restrictions that favor small or minority-owned businesses.

Unlike contractual incentives and allowances, contract requirements can restrict the number of firms willing to construct a particular project. Emission reduction strategies should be designed to maintain free and open competition to the extent practicable.

Retrofit Market. The market for any one technology to reduce emissions from existing construction equipment is relatively small because such equipment varies so greatly in its size, configuration and use, and no one technology will work on more than a subset of the total. The result is a chicken-and-egg problem: construction equipment owners cannot make use of an incentive if suitable technology is not available, but manufacturers may not offer suitable technology until they can see a market large enough to support the cost of doing so. To avoid this problem, incentives have to be left in place long enough, and they have to be inclusive enough, to provide an incentive to the manufacturers, in the first place, to create and offer suitable technologies. A broader and longer-lasting program may be more costly but also increase availability of cost-effective pollution-reducing technologies.

Outreach and Education. Making information available to equipment owners about retrofit alternatives and incentives can be crucial to the success of an incentive program. Owners may need technical assistance in learning how to qualify for an incentive and in evaluating how different alternatives will affect their equipment.

3. Diesel Reduction Incentives

Incentives encourage or promote voluntary efforts to reduce emissions from nonroad construction equipment and would include tax-related incentives, government grants and rebates, low interest loan programs, contractual incentives and allowances, public recognition, non-government financing and fuel supplied by a project owner. Noted in the following are regulatory and contractual requirements that some members of the Clean Diesel and Retrofit Work Group would also like to have considered but other members of the group consider premature to suggest and legally questionable. Everyone agrees that the construction industry faces unique technical and economic challenges in reducing emissions from existing diesel engines and therefore requires a creative approach to retrofit. The remainder of this section describes the most prominent points associated with each incentive as it relates to the construction sector.

Income Tax Incentives. Tax measures that defray part or all the cost of purchasing and installing retrofit technology (e.g., forgiveness, credits and/or accelerated depreciation) have the potential to influence private owners of construction equipment. Some of these measures, however, are useful to only those equipment owners who would otherwise have a tax liability against which to apply the incentive. The Internal Revenue Service has figures showing that only 60% of all corporations in the construction industry in reported net profits in 2001.[30]

Low Interest Loan Programs. These loan programs provide short-term funding for a long-term payoff in diesel emissions reductions. However, loans are effective only to the extent that equipment owners expect that reducing their equipments' diesel emissions will benefit their companies, and therefore, justify the cost of purchasing retrofit technology plus the loan interest. It is questionable whether low interest loans are enough of a financial benefit to motivate equipment owners to voluntarily reduce emissions or whether such loans would be appropriate for emissions reduction activities that do not pay for themselves. However, low interest loan programs could be combined with other incentives.

Contractual Incentives and Allowances. Contractual incentives or allowances are different from contract specifications, which are discussed next. Contract modifications can encourage clean diesel construction projects by providing financial rewards for cleaner practices; however, participation in contract modifications is not guaranteed. Contract modifications can be paired with grants or loans, especially to smaller businesses, to help create a level playing field. Contract modifications do provide the ability to target emission reductions where needed, but must be carefully constructed.

Contract Specifications and/or Requirements. Contract specifications refer to the practice of including provisions related to the use of low emissions equipment and/or fuels in public or private contracts for construction services. Contractual requirements are legally enforceable contract terms and conditions related to the use of low emissions equipment and/or fuels. Such programs have been adopted in Massachusetts, New York and other locations.

Construction companies express concern that contract requirements and regulations could provide a competitive advantage to large, private sector equipment owners with sufficient capital to meet cleaner requirements and would discriminate against smaller businesses that could not afford to retrofit equipment.

Regulatory Requirements. Regulatory requirements provide the opportunity to achieve targeted emissions reductions over a broad geographic area. Some members of the group maintain that regulatory requirements are necessary to achieve maximum health benefits from the construction sector; others maintain that they raise complex legal questions. In addition, other members believe it is unfair to ask the owner of engines and equipment which met emissions standards at the time of purchase bear the cost of further reducing emissions of a compliant engine. While the regulatory process would allow the comparison of costs of compliance with the public health benefits, a concern among construction companies is that regulatory requirements would provide a competitive advantage to larger equipment owners with sufficient capital to finance emission reductions strategies entirely on

---

[30] Internal Revenue Service, U.S. Department of the Treasury, *Table 1: 2001, Corporation Income Tax Returns,* www.irs.gov/pub/irs-soi/01co01as.xls.

their own, and discriminate against smaller firms. Historically, regulatory requirements have not been combined with grants or loans to mitigate these costs, but have been provided for early compliance or for exceeding mandatory requirements.[31]

Public Recognition. Recognition programs are relatively easy to implement and provides equipment owners a critical sense of reward and accomplishment for their initiatives. They are most effective when used in combination with other voluntary incentives. The overall effectiveness is likely to depend on the amount of positive publicity and/or prestige they can generate for equipment owners.

Non-government Financing. This typically involves a private organization (whether nonprofit or profit) that raises capital from various "investors" and then provides funding to equipment owners (whether public or private) to purchase retrofit technology or pursue other emission reduction strategies. The government agency that wants to reduce either its own or others' diesel emission reductions then reimburses the private organization for the funding, providing the organization and/or its "investors" with a financial return in the form of a low-level multi-year payback of the funding. The benefits of this approach are that the government agency desiring the emission reduction project does not need to have on hand the full funding that may be necessary to pursue emission reduction strategies, the equipment owner does not need to incur the expense of pursuing those strategies, and the cost of the retrofit project can be spread over an extended time period. The financing entity assumes the risk for any payments that are not made. This incentive is a particularly good complement to work in combination with other measures as a means of accomplishing diesel emission reductions in an affordable manner without placing undue financial burdens on the equipment owner or government agency desiring the emission reduction benefits.

Fuel Supplied by Project Owner. This incentive shifts the financial burden of purchasing cleaner (e.g., ultra-low sulfur) fuel onto the construction project owner (or other client of the construction contractor). The project owner could pay for the full cost of cleaner fuel, or just the difference in the cost of regular and cleaner fuel. If the project owner paid the full cost of the cleaner fuel, it could require the construction contractor to deduct the cost of the fuel it would otherwise have to purchase from its bid, and at least theoretically, the overall cost of the construction project would still increase by only the incremental cost of the cleaner fuel. However, this arrangement could implicate the terms or conditions of the equipment owner's fuel supply contract, or raise questions about fuel quality and/or equipment warranties.[32]

---

[31] California is the only state thus far to pass regulations requiring the cleanup of diesel trucks and buses, with between 15 and 20% of diesel pollution sources currently regulated. In the next two years, California plans to regulate all sources of diesel pollution, except federally preempted sources like trains and ships.

[32] California is the only state thus far to pass regulations requiring the cleanup of diesel trucks and buses, with between 15 and 20% of diesel pollution sources currently regulated. In the next two years, California plans to regulate all sources of diesel pollution, except federally preempted sources like trains and ships.

As can be seen from the above, each of these measures has both positive and negative attributes. Performance of retrofit programs may be optimized by combining different incentives. Developing, combining and coordinating incentive programs within a region, within a construction project, or across multiple projects within a region is likely to be more effective than attempting to structure one or more incentives independently.

4. Recommendations

The following action items are recommended for EPA to consider in developing a diesel emission retrofit strategy for the construction sector:

Developing Programs, Assistance or Model Language

- Develop and encourage innovative ways to leverage the combination of private financing of investor funds with available government grant funds (including tax incentives, rebates and performance bonuses) to maximize the benefits to equipment owners and minimize the burdens on recipient agencies. Ensure that Federal agencies such as EPA who operate grant and loan programs have adequate resources to successfully administer the programs.

- Provide more opportunities for government grants and rebates to be given to non-profit and/or for-profit entities to avoid the cost and burden to state/local government agencies associated with grant administration and retrofit product acquisition.

- Develop a program of low interest loans to assist state and local governments in increasing the support for funding of retrofit projects.

- Work with the construction industry and government procurement officials to establish model language for contract allowances and incentives and project-specific contract specifications leading to consistent mechanisms for encouraging emission reduction strategies.

- Develop model language for voluntary construction retrofit programs that if implemented by states would qualify for SIP credit and not be calculated as part of the maximum 3 % allowance for voluntary programs. An example of such a program would include the Texas TERP where participation is voluntary but is an enforceable measure in the SIP.

- Investigate and assess operational modifications that have emissions benefit. Then work with the construction industry and government agencies that create construction projects to develop a set of effective and appropriate guidelines for idle reduction, and effective maintenance and repair programs designed to reduce emissions from construction equipment/operation.

- Providing Information, Education and Outreach

- Make available information on, and support relationships with the numerous other grant programs from other Federal and state agencies to help broaden the overall funding pool and leverage available EPA grant funds for retrofit-related projects.

- Develop tools for making good policy decisions regarding reducing emissions from construction equipment. This would include improving the construction industry's emissions inventory and equipment populations and developing a framework for characterizing and quantifying the economic benefits of various approaches to financing the acquisition of retrofit products for the construction industry, and make the information available to state/local/regional government agencies as a tool to guide their decisions on structuring retrofit programs.

- Develop a model process and guidelines that can be used for construction projects to provide rational estimates of emission reductions and related cost effectiveness from use of retrofit products.

- Work with interested stakeholders by establishing ongoing outreach and educational initiatives in the construction sector, including a website (maintained by either EPA or a private sector or industry association organization) targeted to the construction sector.

- Assess and encourage the combination of replacement and repowers with retrofit devices.

Improving the Verification Process

- Accelerate the process for verifying retrofit technologies for use in the construction sector. EPA should evaluate: 1) establishing a special (less burdensome) process for extending the verification for products already verified for on-road applications to nonroad applications, and/or consider 2) establishing a conditional verification with a finite duration (e.g., six months, one year) based on an initial demonstration of technical performance with a requirement for additional technical support to be submitted to obtain full verification status.

- Investigate and where at all practical, incorporate (possibly on an interim status basis) the use of technologies and products that have been approved via the European VERT verification process, as a measure to advance the availability of retrofit products for construction and other nonroad applications.

- Investigate the approval of more labs so that competition among various labs could possibly reduce the cost of testing.

Appendix A. Work Group Members and Organizations

Retrofit Work Group Co-chairs	
Corning, Inc.	Tim Johnson
U.S. Environmental Protection Agency	Gay MacGregor

Ports Sector Co-chairs	
Emisstar	Michael Block
U.S. Environmental Protection Agency	Trish Koman

Construction Sector Co-chairs	
Associated General Contractors of America	Leah Wood Pilconis
U.S. Environmental Protection Agency	Steve Albrink

Freight Sector Co-chairs	
Diesel Technology Forum	Allen Schaeffer
U.S. Environmental Protection Agency	Mitch Greenberg

School Bus Sector Co-Chairs	
National Association of State Directors of Pupil Transportation Services	Charlie Gauthier
U.S. Environmental Protection Agency	Jennifer Keller

Voting Members (Each organization has one vote)	
American Association of Port Authorities	Meredith Martino
American Trucking Association	Glen Kedzie
Associated General Contractors of America	Ken Simonson
BP Global Fuels & Technology/Amoco	Bob Schaefer
California Air Resources Board	Ann Herbert
	Scott Rowland
Caterpillar	Terry Goff
	Patrick Mohrman
Clean Air Action	Ben Henneke
Cummins	Robert Jorgensen
Diesel Technology Forum	
Emisstar	
Emissions Advantage	Bruce Bertelsen
	Tom Timbario
Engine Manufacturers Association	Jed Mandel
	Kevin Kokrda
Environmental Defense	Michael MacLeod
	Janea Scott
Federal Highway Administration	Diane Turchetta
Federal Railroad	Steve Ditmeyer
	Steve Grimm
Fleetguard Emissions Solutions	Jennifer Kain
	Loretta Evans
Georgia Institute of Technology	James Pearson
Infineum	Kevin Poindexter
	Randy Evans

International Truck and Engine Corporation	Peter Reba
John Deere	Glen Chruschiel
	Howard Gerwin/
	Taylor Davis
Lubrizol	Michelle Bellamy/ Kevin Brown
Manufacturers of Emission Controls Association	Dale McKinnon
	Antonio Santos
National Association of Student Transportation	Robin Leeds
National Association of State Directors of Pupil Transportation Services	
National Biodiesel Board	Tom Verry
Natural Gas Vehicle Coalition	Paul Kerkhoven
Natural Resources Defense Council	Rich Kassel
New York Department of Environmental Conservation	Steve Flint
	Thomas Lanni
Northeast States for Coordinated Air Use Management	
Propane Education Research Council	Brian Feehan
Puget Sound	Dennis McLerran
Port of Seattle	Barbara Cole
U.S. Maritime Administration	Danny Gore
	Daniel Yuska
Union of Concerned Scientists	Patricia Monahan
	Michele Robinson
Western Governors' Association	Lee Alter

Other Participating Organizations

Alberici Group
Alcaide & Fay
American Petroleum Institute
Ames Construction
APM Terminals
Carnegie Mellon Department of Mechanical Engineering
Cleaire Advanced Emission Controls, LLC
Clean Diesel Technologies
Coordinating Research Council
DaimlerChrysler
Emissions Technology, Inc.
EPA Region 5
EPA Region 9
Exxon Mobil
Foothills Contracting
Hawn Dredging
Infineum USA LP
Johnson Matthey
Marathon Ashland

National Association of Waterfront Employers
Nett Technologies
New Jersey Department of Environmental Protection
NexGen Fueling
Port of Long Beach
Port Authority of New York & New Jersey
Port of Corpus Christi
Port of Houston Authority
Port of Los Angeles
Virginia International Terminals, Inc.
Propane Education & Research Council
The Accord Group
USCAR
Van Ness Feldman
Volvo/Mack
West Coast Coalition
Williams Brothers Construction

Special Thanks:

Dr. Ralph Appy, Port of Los Angeles
Rick Bayles, Infineum
Dana Blume, Port of Houston Authority
Chuck Carroll, National Association of Waterfront Employers
Dan Demers, Virginia International Terminals, Inc.
Thomas Jelenic, Port of Long Beach
Bob Kanter, Port of Long Beach
Mike Kennedy, AGC
Sarah Kowalski, Port of Corpus Christi
Marty Lassen, Johnson Matthey
Dr. Shokoufe Marashi, Port of Los Angeles
Bob Marcolina, NJ DEP
Joseph Monaco, PANYNJ
Gabe Rozsa, NSTA
Wayne Pighin, APM Terminals
Chuck Salter, Volvo
Tom Timbario, Emissions Advantage
Tod Wickersham, Emisstar
Tay Yoshitani, National Association of Waterfront Employers
Joseph Bachman, EPA
Kathleen Bailey, EPA
Monica Beard-Raymond, EPA
Jim Blubaugh, EPA
Cassie Flynn, EPA
Sally Newstead, EPA
Rosa Shim, EPA
Peter Truitt, EPA
Kuang Wei, EPA
Jennifer Went, EPA

Urszula Mierzio, Johnson Matthey

The workgroup would like to thank EPA's staff for its support as well as EPA's Office of Transportation and Air Quality's Margo Oge and Merrylin Zaw-Mon for helping this process go forward.

Appendix B. Emission Control Technology (ECT) Overview for Ports and Construction Sectors

A. Refuel

Refueling involves substituting existing diesel fuel with cleaner fuels that have been tested and verified by EPA and/or ARB for emissions performance. EPA and ARB currently have verified cleaner fuels for on-road applications but not for nonroad. The following table lists the different types of cleaner fuels that are viable for diesel reduction as well as the benefits and feasibility of their implementation.

Switching to cleaner fuels is one of the most promising of the diesel reduction strategies for the ports and construction sectors because it is a drop-in substitute and ULSD will be widely available when it is required by October 2006 for on-highway applications. Even today, nonroad equipment could be fueled with ULSD on a voluntary basis, reducing prevailing sulfur levels of approximately 3,000 ppm to 15 ppm. ULSD is easily adaptable and does not require equipment changes, or engine replacement or modification. It also reduces SO_2 and PM emissions and enhances retrofit technology, enabling the use of DPFs. ULSD is currently being used by ports across the U.S., as well as for other applications including school and transit buses, and trucks. EPA will require that nonroad diesel fuel sulfur content be limited to 500 ppm in mid 2007 and then to 15 ppm (ULSD) in 2010 for nonroad equipment and for locomotive and marine fuel in 2012.

Table B-1 presents refueling strategies.

Table B-1. Refueling Strategies

Emission Reduction Strategy	Description	HC (%)	PM (%)	NOx (%)	Costs	Nonroad Verified?	Benefits	Issues
Ultra-low sulfur diesel fuel (ULSD) (15 ppm cap) Low sulfur diesel (LSD) (500 ppm cap) ARB on-road diesel (150 ppm cap)	Switching to cleaner diesel fuel with lower sulfur content for PM reduction		5% ULSD (when compared with LSD); 9% ULSD (when compared to unregulated nonroad fuel)		ULSD 3-20+ cents/gal over current on-highway LSD	LSD required in 2007; ULSD in 2010, except marine and locomotive	1. SO_2 and PM reduced 2. Retrofit devices enhanced 3. Lower sulfur fuels have cleaning effect on engines which extends oil change intervals and reduces maintenance costs 4. Does not require equipment changes or modification 5. Relatively easy to adapt 6. Most popular/viable among ports 7. Low sulfur diesel mandated for nonroad, C1, C2 marine, and locomotive in 2007. ULSD required for nonroad (2010), locomotive and marine (2012) 8. Used for marine vessels, dockside equipment, construction equipment, trucks 9. Widely used in demonstration projects	1. Not widely available in certain areas of country 2. Incremental increase in fuel and delivery costs 3. Some reports of fuel pump issues on older engines 4. May need to recondition some older engines
Emulsified diesel	Water and additives mixed with fuel to lower combustion temperatures. Additives prevent water from contacting engine		20-50	5-30	10-20+ cents/gal more	Y	EPA and ARB verified for dockside ports equipment, construction equipment and trucks.	1. May reduce engine power 2. Increases fuel consumption 3. Availability 4. Cold weather operation may be compromised. 5. Special fuel storage requirements (needs periodic in-tank mixing) 6. Early concerns with compromised engine durability

Table B-1. Refueling Strategies

Emission Reduction Strategy	Description	HC (%)	PM (%)	NOx (%)	Costs	Nonroad Verified?	Benefits	Issues
Biodiesel	Renewable fuel (meeting ASTM spec 6751) made from vegetable oils/animal fats	20-30	2-10	0-2 increase (with B20)	5-20 cents/gal more	Y	1. Reduces PM, CO, HC 2. Various blends available: B20 is 20% biodiesel, 80% diesel. 3. Verified for trucks	1. May increase NOx 2. B100 not recommended for cold weather operation 3. Needs to meet ASTM specs 4. Care needed for transport and storage 5. Engine manufacturers typical warrant only up to 5% biodiesel blends. 6. Currently a large variability in biodiesel quality.

B. Retrofit

Retrofitting is a term used to describe the installation of emission control technologies on in-use equipment and vehicles to reduce PM, NOx, and other pollutants. These technologies have been rigorously verified by EPA and ARB to reduce diesel emissions. DOCs and DPFs are widely used across the country in many different applications.

Table B-2 lists and describes the available retrofit technology.

Table B-2. Available Retrofit Technology for the Ports and Construction Sector

Emission Reduction Strategy	Description	HC (%)	PM (%)	NOx (%)	Costs [33]	Fuel Requirements	On-road/ Nonroad Verified	Benefits	Issues
Diesel Particulate Filters (DPFs)	A passive filter-honeycomb device placed within the exhaust stream that physically traps and oxidizes PM; can be combined with NOx retrofit technologies for NOx reductions.	50-95	85+	---	$5,000-$10,000	ULSD	Y/Y	1. Reduce HC, CO, and PM (including ultrafine particulate) 2. Reduce smoke and odor 3. Passive filters verified for dockside equipment and construction equipment, trucks 4. Typically a direct replacement for the current muffler	1. Passive filters require specific exhaust temperature profiles and appropriate duty cycles to work properly. 2. Filters require maintenance depending on duty cycle 3. Passive filters must be operated with ULSD
Flow-Through Filters	A flow-through filter does not physically "trap" and accumulate PM, but instead exhaust flows through a medium that has a high density of torturous flow channels, giving rise to turbulent flow conditions	50-95	30-60+	---	$5,000-$7,000	500 ppm sulfur; some may require ULSD	Y/N	1. Reduce HC, CO, and PM 2. Reduce smoke and odor 3. No maintenance when applied per guidelines 4. May be more applicable to nonroad engines than DPFs because much less likely to plug under unfavorable conditions	1. Lower PM control compared to DPF 2. May have application guidelines around operating temperatures

---

[33] Costs are based on on-road experience and are expected to be in the same range for similarly-sized nonroad engines.

Table B-2. Available Retrofit Technology for the Ports and Construction Sector

Emission Reduction Strategy	Description	HC (%)	PM (%)	NOx (%)	Costs [33]	Fuel Requirements	On-road/ Nonroad Verified	Benefits	Issues
Diesel Oxidation Catalysts (DOCs)	Devices that oxidize PM and HC. Can be bolted onto exhaust, or direct muffler replacement; can be coupled with other retrofit technologies for additional PM and/or NOx reductions.	50-90	25-50	---	$500-$2,000	LSD	Y/Y	1. Reduce HC, CO, and PM 2. Reduce odor 3. Established record in both highway sector and nonroad applications (in use for over 35 years) 4. Requires no continual maintenance 5. Verified for dockside equipment and construction equipment, trucks	1. Works better with lower sulfur diesel (less than 350 ppm); works best with ULSD
Selective Catalytic Reduction (SCR)	System injects urea (or ammonia) into exhaust stream and reacts this mixture over a catalyst to reduce NOx emissions. Can be used in conjunction with DOC or DPF.	80	20-30 variable see benefits	~80	$12,000 to $15,000 (w/DOC) $15,000 to $20,000 (w/ DPF)	LSD preferred	Y/Y	1. High NOx reductions 2. PM reduced additional 25% with a DOC; and up to 85% with a DPF and ULSD	1. Requires periodic refilling of ammonia/urea tank 2. Requires urea supply infrastructure 3. Commonly used in stationary applications (power plants), recently adapted to vehicles 4. Potential safety concerns for ammonia reductant
Lean NOx Catalysts (LNCs)	Injects diesel fuel into exhaust stream and reacts this mixture over a catalyst to reduce NOx emissions. Verified LNCs always paired with DOC or DPF.	---	10-20 (w/ DOC); 85+ (w/ DPF)	5-30	$15,000-$20,000	LSD preferred	Y/N	Reduce NOx	Can increase fuel usage by 5-7%

Table B-2. Available Retrofit Technology for the Ports and Construction Sector

Emission Reduction Strategy	Description	HC (%)	PM (%)	NOx (%)	Costs [33]	Fuel Requirements	On-road/ Nonroad Verified	Benefits	Issues
Exhaust Gas Recirculation (EGR) with a DPF	Recirculates engine exhaust back to engine to cool peak combustion temperatures and reduce NOx; paired with DPF	~80%	85+	40-50	$18,000-$20,000	ULSD	Y/N	Reduces NOx and PM	The feasibility of low-pressure EGR is more of an issue with nonroad equipment than on-road equipment (i.e., more difficult to cool the exhaust).
Closed Crankcase Ventilation (CCV)	Directs engine's blow-by gases (NOx, HC, and toxics) to intake system for recombustion instead of polluting environment. PM collected in filter, removed from crankcase vapors; can be paired with other emission control strategies	---	5-10	---	$450-$700	LSD	Y/Y	1. Reduces PM 2. Effective strategy for reducing in-cabin PM exposure	1. Requires periodic change of a disposable filter (i.e., at every oil change) with some designs 2. Engine compartment space constraints for some applications
Idle Reduction Technologies	Devices prevent operators from idling for long periods. Can include shut-off devices or auxiliary power units (APUs).	Varies			Varies depending upon technology	None	N/N	1. Reduce NOx, PM, CO, HC 2. Some devices pay for themselves in a short time through fuel savings 3. Can save wear and tear on engines and reduce maintenance costs 4. Can be used in for locomotives	1. Additional cost and complexity 2. Additional weight of auxiliary power unit (APU) 3. Potential space constraints

C. Operational Strategies

Operational strategies can be used to reduce diesel emissions. These strategies are cost-effective and make good business sense by maximizing efficient use of port equipment and vehicles while optimizing the flow of cargo in and out of the port. With new Homeland Security requirements, some port authorities are looking for opportunities to reduce emissions while enhancing security and modernizing information technology (IT).

Table B-3 presents operational strategies available for ports.

Table B-3. Operational Strategies for the Ports and Construction Sectors

Emission Reduction Strategy	Description	Emission Reductions (Benefits)	Estimated Costs	Benefits	Issues
Gate efficiencies	Gate appointments systems with web-based access allow carriers to pickup/drop off when they want. WebAccess customers update and view data on 24/7 basis			1. Truck idle reduction 2. Allows 240 gate transactions/hr vs. manually (improve 84% efficiency) 3. gate process expedited 4. eliminates trouble transactions 5. Allows customers to "pre-gate" or be alerted for advance pickup/drop off. 6. Homeland security gives money for pilot projects 7. Port of New Orleans only port that mandated gate entry management system 8. Georgia Ports case study (30% reduction in turn times, 3020 gallons fuel saved/day, half ton NOx/ day, 33 tons CO_2 reduced/peak day) 9. Port Houston building a pre-check gate facility that reduces processing times for entering trucks from 22 min to 6 min. 10. Improves port efficiency 11. Improves port security	1. Education 2. No consistent methodology for calculating turn times that considers truck wait times and idling outside the gate(s).
E-Modal logistics, scheduling and appointments	Use of IT for scheduling operations such as cargo pickup/drop-off to improve port efficiency			1. Efficiency in flow of goods 2. Reduces trouble transactions 3. Reduces idling and congestion 4. Currently done in New Orleans	1. Cost of software 2. Education of logistics and software

Table B-3. Operational Strategies for the Ports and Construction Sectors

Emission Reduction Strategy	Description	Emission Reductions (Benefits)	Estimated Costs	Benefits	Issues
Expanded hours or incentives for off-peak operation to avoid lines	Extend terminal gate hours beyond regular schedule			1. Reduces queuing, truck idling, traffic congestion 2. Increase flow and efficiency 3. Not working during peak ozone hrs	1. Need for labor agreements for expanded hours 2. Freight recipients must be willing to extend their dock operating hours. 3. Incentives for after hours workers
Cold Ironing (Onshore Power)	Uses electric shore side power at berth rather than running auxiliary diesel engines Strategy is most effective for ports and vessels that have long hotelling times, multiple annual vessel calls, and high auxiliary power needs.	Oceangoing vessels account for 32% of all marine vessel NOx emissions at Houston and 20% at POLA		1. NOx, SOx, PM, CO, HC reduced 2. Targets oceangoing vessels (which when hotelled, account for 32% of all marine vessel NOx emissions at Houston and 20% at POLA) 3. Already used by several ports (POLA). Successful for Princess cruise ships in Alaska, Seattle, most Navy terminals	1. Requires an infrastructure investment (electricity supply) 2. Oceangoing vessels must be retrofitted to be able to receive shore power at the port (allowing aux. engine shutoff) 3. High cost
Substitute Electric Power for Diesel Power: Electric Dredging and Electric Cranes	Substitutes diesel powered dredging equipment or cranes for electrically powered		Estimates of $1M over diesel-powered cranes	1. Local NOx, PM, HC, CO 2. Already used by several ports (PANYNJ).	High cost
Voluntary Reduced Idling[34]	Engine idling for extended periods (usually for heating, air conditioning) is unnecessary. Some truck engines equipped with automatic idle shut-off devices.			1. Reduces emissions 2. Saves fuel and maintenance cost from wear and tear of engine 3. APU (auxiliary power unit) can be used to provide power during idling. APUs produces far less emissions of PM, NOx, than diesel. 4. Applies to trucks, locomotives, equipment	1. Requires training and education to help encourage equipment operators to shut down engine 2. Incentives to encourage voluntary anti-idling such as air conditioned rooms or heated rooms

---

[34] Idle reduction has been the only operational/technology strategy identified so far by the Construction subgroup

Table B-3. Operational Strategies for the Ports and Construction Sectors

Emission Reduction Strategy	Description	Emission Reductions (Benefits)	Estimated Costs	Benefits	Issues
Voluntary Marine Vessel Speed Limits	Emissions typically increase with speed, so enacting speed limits can reduce emissions. Port can implement a "reduced speed zone"			1. Reduces NOx, PM, HC, CO 2. Targets ocean-going vessels, tugboats, ferries 3. POLA, POLB established Voluntary Commercial Ship Speed Reduction Program which urges vessels to travel below 12 knots within 20 miles of coast	1. Limited to a certain distance from port 2. Applicability depends on port-specific configuration 3. May need to offer incentives for voluntary speed limits or enforcement of required slow speed zones
TWIC (Transportation Worker Identification Credential)	Identity credential for un-escorted physical access to secure areas and cyber systems			1. Improves security (Homeland security requires identification and screening of employees) reduce need for multiple credentials 2. reduce need for multiple credentials 3. Can reduce idling for truckers in line (instead of a person having to look at each ID)	
Container Management using information technologies (IT) to improve stacking practices, container tracking, direct intermodal transfers (cargo moved directly from ship to rail), homeland security changes	Minimizing container moves reduces emissions of cargo handling equipment			1. Tracking and virtual container yards using IT enables more efficient movement of cargo 2. For dockside equipment 3. Reduces NOx, PM, HC, CO 4. Fuel savings and better equipment maintenance 5. Involves security 6. Less time wasted sifting through and looking for containers	1. Implementing IT 2. Cost

Table B-3. Operational Strategies for the Ports and Construction Sectors

Emission Reduction Strategy	Description	Emission Reductions (Benefits)	Estimated Costs	Benefits	Issues
Substitute rail (e.g. "on-dock"), barge or short sea shipping instead of trucking where feasible	Compared to trucking, barge and rail emissions can be low if the barge or locomotive is new or has been retrofitted or repowered. Reduces congestion. Typically trucks used to transfer containers between port and intermodal rail facility.			1. Reduce NOx, PM, CO, HC 2. more cargo can fit on rail, barge, or ships 3. Ports served by railroad can have containers moved directly from marine vessels to rail, eliminating movement of on-road trucks. Trucks tend to be oldest and highest polluting in operation. Use of on-dock rail is effective in reducing congestion 4. Example at Port of NY & NJ and Port of Seattle designed with on-dock rail	

D. Repair/Rebuild

Engines that are properly maintained and tuned perform better and typically emit less pollution than engines that are not properly maintained. Rebuilding an engine as a strategy for emissions reduction can also significantly lower emissions, run more efficiently, and be cost-effective for high value equipment. Proper maintenance or rebuilding lowers emissions by burning fuel more efficiently and can reduce operation costs and extend engine life.

The following is a list of maintenance issues to consider:

- Restricted air filters
- Improper engine timing
- Clogged, worn or mismatched fuel injectors
- Faulty fuel injection pumps
- Defective or misadjusted puff limiters
- Low air box pressure
- Improperly adjusted valve lash or governors
- Malfunctioning turbo chargers and after-coolers
- Maladjusted fuel racks
- Defective air fuel controllers
- Poor fuel quality
- Improper driving gear
- Air intake manifold leaks

E. Repower

Repower is a term used to describe replacing an older engine with a newer cleaner engine or replacing a diesel engine with one that can use alternative fuels. Table B-4 shows the different techniques for repowering engines.

Table B-4. Repower Options for the Ports and Construction Sectors

Emission Reduction Strategy	Description	Emission Reductions (Benefits)	Costs	Fuel Requirements	Benefits	Issues
Repower with newer, cleaner diesel engine	Removing an older engine and replacing with a newer cleaner engine	Variable, depending upon "Tier level" of old engine cf "Tier level" of new engine		Up to 2008, diesel quality fuel independent; for 2008+, ULSD	1. Reduces NOx, PM, HC, CO 2. For marine vessels (aux. and propulsion engines), construction and dockside equipment, trucks, rail	1. May pose technical issues – need to consult original engine or equipment manufacturer
Repower with alternatively fueled engine	Remove older diesel engine and replace with an alternatively fueled engine	Variable, depending upon "Tier level" of old engine cf spec (fuel type, etc.) of new engine		Alt fuel	1. Reduces PM &/or NOx 2. For marine vessels, construction, dockside, trucks, rail	1. Cost for fuel and conversion 2. May require fuel infrastructure (e.g., CNG)
Replace a nonroad engine with hwy engine manufactured to stricter standards	Substitute a highway engine for a comparable model year or older nonroad engine			Up to 2007, LSD; for 2007+, ULSD	1. Reduce NOx, PM, HC, CO 2. For yard tractors and cargo handling equipment that have duty cycles similar to highway engines are good candidates	1. Requires highway grade fuel

F. Replace

As the emissions standards change, newly manufactured engines must meet new emissions requirements. Voluntarily replacing older diesel equipment prior to the end of their operational life with diesel equipment that meets tougher emissions requirements is a viable and often cost-effective strategy for cleaner air. Replacing also involves the scrapping of the old engine/equipment to ensure it does not reappear in the marketplace in another location and continue to contribute to excess diesel emissions.

Table B-5 presents options for replacing diesel equipment.

Table B-5. Equipment Replacement Options for the Ports and Construction Sectors

Emission Reduction Strategy	Description	Emission Reductions (Benefits)	Costs	Fuel Requirements	Benefits	Issues
Replacing older diesel equipment with newer diesel equipment	Replacing older vessels, equipment, trucks and switchers with ones that are newer and cleaner.			Up to 2008, diesel fuel quality independent; for 2008+, ULSD	1. Typically, NOx, PM, HC, CO reduced 2. Marine vessels, construction and dockside equipment, trucks and rail 3. Turnover of equipment allows for replacement 4. Most cost effective when uncontrolled engines are replaced such as pre-1984 trucks or pre-1996 nonroad equipment 5. Typically benefits in fuel efficiency, reliability, warranty and maintenance costs.	Cost
Replacing nonroad equipment with new models equipped with certified on-road engines	Highway equipment is cleaner than nonroad equipment in comparable model years. Therefore specifying highway engines in yard trucks and applicable landside equipment reduces emissions.				1. Typically, NOx, PM, HC, CO reduced 2. Dockside equipment such as yard tractors that have duty cycles similar to highway engines. 3. Can save money through significant (up to 20%) fuel savings and come with additional safety features. 4. Port of NY & NJ container terminal tenants are doing this	Applies to specific conditions
Replacing diesel equipment with electric, hybrid or alternative fuel equipment (LNG, CNG, propane)	Can replace diesels with those with utilizing hybrid technology or alternative fuels.				1. Typically, NOx, PM, HC, CO reduced 2. Marine vessels, construction and dockside equipment, trucks and rail 3. Examples include hybrid switcher locomotives, electric cranes, LNG or LPG yard tractors, forklifts or loaders.	1. Natural gas replacements may require fueling infrastructure. 2. Cost for fuel and hybrid

Appendix C. Description of Verification Programs

The objective of the Voluntary Diesel Retrofit Program Verification is to introduce verified technologies to the market as cost effectively as possible, while providing customers with confidence that verified technologies will provide emission reductions as advertised. EPA and CARB's verification process evaluates the emission reduction performance of retrofit technologies, including their durability, and identify engine operating criteria and conditions that must exist for these technologies to achieve those reductions.

EPA and ARB signed a Memorandum of Agreement (MOA) for the coordination and reciprocity in Diesel Retrofit Device Verification. The MOA establishes reciprocity in verifications of hardware or device-based retrofits, and further reinforces EPA's and ARB's commitment to cooperate on the evaluation of retrofit technologies. This agreement commits EPA and ARB to work toward accepting particulate matter (PM) and oxides of nitrogen (NOx) verification levels assigned by the other's verification programs. Additionally, as retrofit manufacturers initiate and conduct in-use testing, EPA and ARB agreed to coordinate this testing so data generated may satisfy the requirements of each program. This MOA is intended to expedite the verification and introduction of innovative emission reduction technologies. Additionally, this MOA should reduce the effort needed for retrofit technology manufacturers to complete verification.

Information about CARB's Verification Program and its list of verified technologies can be found at the ARB verification page at http://www.arb.ca.gov/diesel/verdev/verdev.htm. Information about EPA's Verification program and its list of verified technologies can be found on EPA's verification page at http://www.epa.gov/otaq/retrofit/retroverifiedlist.htm

Table C-1 presents all the diesel retrofit products that have been approved for use in off-road engine retrofit programs.

Table C-1. Verified Off-Road Technologies

Company	Product Name/ Technology Type	Applications	Example Equipment Types	PM Reduction (%)	NOx Reduction (%)	Fuel Type
Lubrizol Engine Control Systems	Lubrizol PuriNOx / Water Emulsion (Alternative) Fuel ○	Heavy Duty, 2 & 4 Cycle engines	All off-road & highway diesel engines	16.8 to 23.3	17 to 20.2	Emulsified fuel with 2D having 500 ppm sulfur
Lubrizol Engine Control Systems	Lubrizol AZ Purimuffler and Purifier / Diesel Oxidation Catalyst (DOC) + PuriNOx ●	Certain 1996-2002 off-road port, railway yard, and other intermodal + freight handling operation equipment	Includes Case, Komatsu, Cummins, & International engines	50 (Level 2)	20	Emulsified fuel
Lubrizol Engine Control Systems	Lubrizol ECS AZ Purifier and Purimuffler / DOC ●	Certain 1996-2002 off-road port, railway yard, and other intermodal + freight handling operation equipment	Includes Cummins & International engines	25 (Level 1)	NA	15 ppm sulfur diesel
Lubrizol Engine Control Systems	Lubrizol ECS Unikat Combifilter / Diesel Particulate Filter (DPF) ●	Certain 1996-2004 off-road applications used in construction, material and cargo handling equipment	Includes most off-road engines by most manufacturers	85 (Level 3)	NA	CARB diesel or 15 ppm sulfur diesel
Donaldson Company	Donaldson / Series 6000 DOC + Crankcase Filter ●	Certain 1996-2003 off-road engines used in yard tractors, large lift trucks, top picks, side picks, and gantry cranes	Includes turbocharged engines from 150 – 600 hp by Case, CAT, Cummins, DDC & Komatsu	25 (Level 1)	NA	CARB diesel or 15 ppm sulfur diesel

Table C-1. Verified Off-Road Technologies

Company	Product Name/ Technology Type	Applications	Example Equipment Types	PM Reduction (%)	NOx Reduction (%)	Fuel Type
CleanAIR Systems	CleanAIR Systems / DPF ●	Certain 1996-2003 off-road engines used in stationary emergency generators.	Includes most manufacturers of stationary emergency generators	85 (Level 3)	NA	15 ppm sulfur diesel
Extengine Transport Systems	Extengine - Advanced Diesel Emission Control System (ADEC) / DOC + Selective Catalytic Reduction (SCR) ●	Certain 1991-1995 Cummins 5.9-liter, 150 to 200 HP off-road engines used in excavators, dozers, loaders, and utility tractor rigs	Includes only Cummins 5.9 engines	25 (Level 1)	80	CARB diesel or 15 ppm sulfur diesel
Caterpillar, Inc.	Diesel Particulate Filter ○	Nonroad, 4-cycle, non-EGR equipped, model year 1996-2005, turbocharged engines with power ratings between 174.2 to 301.5 Horsepower	Certain Caterpillar off-road engines	89	NA	15 ppm sulfur diesel

○ EPA Verified Technologies are listed and explained at: http://www.epa.gov/otaq/retrofit/retroverifiedlist.htm

● CARB Verified Technologies are listed and explained at:: http://www.arb.ca.gov/diesel/verdev/verdev.htm